AF581765

Biomedical Image Processing and Classification

Biomedical Image Processing and Classification

Editor

Luca Mesin

MDPI • Basel • Beijing • Wuhan • Barcelona • Belgrade • Manchester • Tokyo • Cluj • Tianjin

Editor
Luca Mesin
Polytechnic University of Turin
Italy

Editorial Office
MDPI
St. Alban-Anlage 66
4052 Basel, Switzerland

This is a reprint of articles from the Special Issue published online in the open access journal *Electronics* (ISSN 2079-9292) (available at: https://www.mdpi.com/journal/electronics/special_issues/bioimg_classification).

For citation purposes, cite each article independently as indicated on the article page online and as indicated below:

LastName, A.A.; LastName, B.B.; LastName, C.C. Article Title. *Journal Name* **Year**, *Volume Number*, Page Range.

ISBN 978-3-0365-0346-2 (Hbk)
ISBN 978-3-0365-0347-9 (PDF)

Contents

About the Editor

Luca Mesin (Ph.D.) is currently employed at the Department of Electronics and Telecommunications, Polytechnic University of Turin, 10129 Turin, Italy. He graduated with a degree in Electronics Engineering in 1999, and he received a Ph.D. in Applied Mathematics in 2003 from Politecnico di Torino, Italy. His present position is that of Associate Professor in Biomedical Engineering and he is also the supervisor of the Mathematical Biology and Physiology group of the Department of Electronics and Telecommunications, Politecnico di Torino. His main research activities concern biomedical signal/image processing and classification, biophysical modeling and human–machine interactions.

Editorial

Biomedical Image Processing and Classification

Luca Mesin

Mathematical Biology and Physiology, Department of Electronics and Telecommunications, Politecnico di Torino, 10129 Turin, Italy; luca.mesin@polito.it; Tel.: +39-0110904085

Received: 30 November 2020; Accepted: 14 December 2020; Published: 1 January 2021

1. Introduction

Biomedical image processing is an interdisciplinary field [1] that spreads its foundations throughout a variety of disciplines, including electronic engineering, computer science, physics, mathematics, physiology, and medicine. Several imaging techniques have been developed [2], providing many approaches to the study of the body, including X-rays for computed tomography, ultrasounds, magnetic resonance, radioactive pharmaceuticals used in nuclear medicine (for positron emission tomography and single-photon emission computed tomography), elastography, functional near-infrared spectroscopy, endoscopy, photoacoustic imaging, and thermography. Even bioelectric sensors, when using high-density systems sampling a two dimensional surface (e.g., in electroencephalography or electromyography [3]), can provide data that can be studied by image processing methods. Biomedical image processing is finding an increasing number of important applications, for example, to make image segmentation of an organ to study its internal structure and to support the diagnosis of a disease or the selection of a treatment [4].

Classification theory is another well developed field of research [5] connected to machine learning, which is an important branch of artificial intelligence. Different problems have been addressed, from the supervised identification of a map relating input features to a desired output, to the exploration of data by unsupervised learning (cluster analysis, data mining) or online training through experience. The estimation of informative features and their further processing (by feature generation) and selection (either by filtering or with approaches wrapped to the classifier) are important steps, both to improve classification performance (avoiding overfitting) and to investigate the information provided by candidate features to the output of interest. Excellent results have also been recently documented by deep learning approaches [6], in which optimal features are automatically extracted in deep layers on the basis of training examples and then used for classification.

When classification methods are associated with image processing, computer-aided diagnosis (CAD) systems can be developed, e.g., for the identification of diseased tissues [7] or a specific lesion or malformation [4]. These results indicate interesting future prospects in supporting the diagnosis of diseases [8].

2. This Special Issue

The present issue consists of six papers on a few topics in the wide range of research fields covered by biomedical image processing and classification.

In [9], the authors have proposed a CAD system for identification and assessment of glomeruli from kidney tissue slides. Their approach is based on deep learning, exploiting convolutional neural network (CNN) architectures tailored for the semantic segmentation task. The obtained results are promising, as also stated by expert pathologists. Moreover, the proposed system can easily be integrated into the existing pathologists' workflow thanks to an XML interface with Aperio ImageScope [10].

With the recent advances of techniques in digitalized scanning, tissue histopathology slides can be stored in the form of digital images [11]. In recent years, many efforts have been devoted to developing automated classification and segmentation techniques with the aim of improving accuracy and efficiency in digital pathology [12]. In kidney transplantations, pathologists evaluate the architecture of renal structures to assess the nephron status. An accurate evaluation of vascular and stromal injury is crucial for determining kidney acceptance, which is currently based on the pathologists' histological evaluations on renal biopsies in addition to clinical data. In this context, automated algorithms may offer crucial support to histopathological image analysis. An example is given in this Special Issue [13].

Although the performance of a machine learning algorithm depends on the amount of available data, few studies have explored the minimal amount of data required to train a CNN in medical deep learning or the possibility of having scarce annotations [14]. An innovative contribution is given in this Special Issue [15]. The paper explores the minimum number of patients required to train a U-Net that accurately segments the prostate on T2-weighted MRI images. A U-Net was trained on patient numbers that ranged from 8 to 320 and its performance was measured. The Dice score significantly increased from training sizes of 8 to 120 patients and then plateaued with minimal improvement after 160 cases. This study suggests that modest dataset sizes could be sufficient to segment other organs effectively as well.

The correlation between conjunctival pallor (on physical examinations) and anemia paved the way for new non-invasive methods for monitoring and identifying the potential risks of this important pathology. A critical research challenge for this task is represented by designing a reliable automated segmentation procedure for the eyelid conjunctiva. A graph partitioning segmentation approach is proposed in [16], exploiting normalized cuts for perceptual grouping, thereby introducing a bias towards spectrophotometry features of hemoglobin. The segmentation task has been further investigated by a subsequent work, proposing a deep-learning-based approach involving a deconvolutional neural network [17]. The overall pipeline for building a reliable estimator is composed of several smaller tasks having multiple research challenges [18,19]. For instance, starting from the digital image capturing phase, the process is affected by heterogeneous ambient lighting conditions and intrinsic color balancing techniques by the device [20].

An efficient framework for enhancing and segmenting brain MRIs to identify a tumor is discussed in [21]. The hybridized fuzzy clustering and distance regularized level set (DRLS) technique effectively extracted the region of interest (ROI) in the brain slices. For identifying the ROI, fuzzy clustering was employed by selecting the number of clusters k, validated using the silhouette metric. In post-processing, the ROI mining techniques, marker controlled watershed segmentation, seed region growing and DRLS were adopted to extract the anomalous section from the segmented objects [22,23]. Tumor volume computation and 3D-modeling of the clinical dataset abnormalities were performed using the physical spacing metadata available in the header of the DICOM images considered. This can help physicists locate the tumor and determine other information (e.g., size and shape) during initial diagnosis, and thereby the process of treating the tumor may be enhanced.

Finally, one paper in this Special Issue has addressed the problem of identifying the volume status of patients [24]. The method was developed within a long-standing research activity on the automated investigation of the pulsatility of the inferior vena cava (IVC) from ultrasound measurements. The clinical approach is based on the subjective choice of a fixed direction along which to investigate IVC pulsations. However, the vein may have a complicated shape and show respirophasic movements, which introduce uncertainties into the clinical evaluation. Two automated methods have been introduced to delineate the IVC edges along sections either transverse or longitudinal to the blood vessel [25–27]. Preliminary results have shown the importance of using these automated methods to obtain more repeatable, reliable, and accurate information on IVC pulsatility than when using subjective clinical methods [28–31]. In this Special Issue, the two views are used to extract features that, integrated by a classification algorithm, can result in improved performance in diagnosing the volemic status of patients [24].

3. Future Perspectives

The research fields of biomedical image processing and classification have reached high levels of insight. Their integration into CAD systems can greatly contribute to supporting medical doctors to refine their clinical picture. In the near future, further growth in contributions to this field is expected; for example, taking advantage of increasing digitalization, deep learning has the potential to provide efficient solutions to many medical problems.

However, the real challenge is to bring an increasing number of systems into the hands of doctors, so that they can be applied to patients. This requires leaving the laboratory, engineering the systems, certifying the products, and identifying the correct target market that can accommodate the new devices and allow adequate support for these activities. In order to speed up this innovation process, the collaboration between researchers, institutions, funders, and entrepreneurs is always more important. The "do-it-all-yourself" approach only makes sense in a world of scarce external knowledge, but today knowledge is spread as it has never been before. Thus, in order to improve the wellness of the whole community [32], a dynamic environment in which new high-impact solutions can be created will be able to grow only if there is collaboration among organizations.

Funding: This research was carried out as part of the project "Method and apparatus to characterize non-invasive images containing venous blood vessels", funded through the PoC Instrument initiative, implemented by LINKS, with the support of LIFFT, with funds from Campagnia di San Paolo.

Acknowledgments: I would like to thank all authors who contributed to this Special Issue, the reviewers for their help in refining the papers and the MDPI editorial board and staff for the opportunity to be guest-editor.

Conflicts of Interest: The author declares no conflict of interest.

References

1. Deserno, T.M. *Biomedical Image Processing*; Springe: New York, NY, USA, 2011.
2. Maier, A.; Steidl, S.; Christlein, V.; Hornegger, J. *Medical Imaging Systems: An Introductory Guide, New York*; Springer: Berlin, Germany, 2018; Volume 11111.
3. Merletti, R.; Muceli, S. Tutorial. Surface EMG detection in space and time: Best practices. *J. Electromyogr. Kinesiol.* **2019**, *49*, 102363. [CrossRef] [PubMed]
4. Mesin, L.; Mokabberi, F.; Carlino, C.F. Automated Morphological Measurements of Brain Structures and Identification of Optimal Surgical Intervention for Chiari I Malformation. *IEEE J. Biomed. Health Inform.* **2020**, *24*, 3144–3153. [CrossRef] [PubMed]
5. Theodoridis, S.; Koutroumbas, K. *Pattern Recognition*; Academic Press: Cambridge, MA, USA, 2008.
6. Bevilacqua, V.; Brunetti, A.; Guerriero, A.; Trotta, G.F.; Telegrafo, M.; Moschetta, M. A performance comparison between shallow and deeper neural networks supervised classification of tomosynthesis breast lesions images. *Cogn. Syst. Res.* **2019**, *53*, 3–19. [CrossRef]
7. Tiwari, A.; Srivastava, S.; Pant, M. Brain tumor segmentation and classification from magnetic resonance images: Review of selected methods from 2014 to 2019. *Pattern Recognit. Lett.* **2020**, *131*, 244–260. [CrossRef]
8. Yanase, J.; Triantaphyllou, E. A systematic survey of computer-aided diagnosis in medicine: Past and present developments. *Expert Syst. Appl.* **2019**, *138*, 112821. [CrossRef]
9. Altini, N.; Cascarano, G.; Brunetti, A.; Marino, F.; Rocchetti, M.; Matino, S.; Venere, U.; Rossini, M.; Pesce, F.; Gesualdo, L.; et al. Semantic Segmentation Framework for Glomeruli Detection and Classification in Kidney Histological Sections. *Electronics* **2020**, *9*, 503. [CrossRef]
10. Altini, N.; Cascarano, G.D.; Brunetti, A.; De Feudis, I.; Buongiorno, D.; Rossini, M.; Pesce, F.; Gesualdo, L.; Bevilacqua, V. A deep learning instance segmentation approach for global glomerulosclerosis assessment in donor kidney biopsies. *Electronics* **2020**, *9*, 1768. [CrossRef]
11. Chen, H.; Qi, X.; Yu, L.; Dou, Q.; Qin, J.; Heng, P.-A. DCAN: Deep contour-aware networks for object instance segmentation from histology images. *Med. Image Anal.* **2017**, *36*, 135–146. [CrossRef]
12. Janowczyk, A.; Madabhushi, A. Deep learning for digital pathology image analysis: A comprehensive tutorial with selected use cases. *J. Pathol. Inform.* **2016**, *7*, 29. [CrossRef]

13. Salvi, M.; Mogetta, A.; Meiburger, K.; Gambella, A.; Molinaro, L.; Barreca, A.; Papotti, M.; Molinari, F. Karpinski Score under Digital Investigation: A Fully Automated Segmentation Algorithm to Identify Vascular and Stromal Injury of Donors' Kidneys. *Electronics* **2020**, *9*, 1644. [CrossRef]
14. Tajbakhsh, N.; Jeyaseelan, L.; Li, Q.; Chiang, J.N.; Wu, Z.; Ding, X. Embracing imperfect datasets: A review of deep learning solutions for medical image segmentation. *Med. Image Anal.* **2020**, *63*, 101693. [CrossRef] [PubMed]
15. Bardis, M.; Houshyar, R.; Chantaduly, C.; Ushinsky, A.; Glavis-Bloom, J.; Shaver, M.; Chow, D.; Uchio, E.; Chang, P. Deep Learning with Limited Data: Organ Segmentation Performance by U-Net. *Electronics* **2020**, *9*, 1199. [CrossRef]
16. Dimauro, G.; Simone, L. Novel Biased Normalized Cuts Approach for the Automatic Segmentation of the Conjunctiva. *Electronics* **2020**, *9*, 997. [CrossRef]
17. Kasiviswanathan, S.; Bai Vijayan, T.; Simone, L.; Dimauro, G. Semantic Segmentation of Conjunctiva Region for Non-Invasive Anemia Detection Applications. *Electronics* **2020**, *9*, 1309. [CrossRef]
18. Dimauro, G.; De Ruvo, S.; Di Terlizzi, F.; Ruggieri, A.; Volpe, V.; Colizzi, L.; Girardi, F. Estimate of Anemia with New Non-Invasive Systems—A Moment of Reflection. *Electronics* **2020**, *9*, 780. [CrossRef]
19. Dimauro, G.; Caivano, D.; Di Pilato, P.; Dipalma, A.; Camporeale, M.G. A Systematic Mapping Study on Research in Anemia Assessment with Non-Invasive Devices. *Appl. Sci.* **2020**, *10*, 4804. [CrossRef]
20. Dimauro, G.; Guarini, A.; Caivano, D.; Girardi, F.; Pasciolla, C.; Iacobazzi, A. Detecting Clinical Signs of Anaemia from Digital Images of the Palpebral Conjunctiva. *IEEE Access* **2019**, *7*, 113488–113498. [CrossRef]
21. Kanniappan, S.; Samiayya, D.; Vincent P M, D.; Srinivasan, K.; Jayakody, D.; Reina, D.; Inoue, A. An Efficient Hybrid Fuzzy-Clustering Driven 3D-Modeling of Magnetic Resonance Imagery for Enhanced Brain Tumor Diagnosis. *Electronics* **2020**, *9*, 475. [CrossRef]
22. Srinivasan, K.; Gowthaman, T.; Nema, A. Application of structural group sparsity recovery model for brain MRI. In Proceedings of the SPIE 10806, Tenth International Conference on Digital Image Processing, Shanghai, China, 11–14 May 2018; p. 108065H.
23. Srinivasan, K.; Ankur, A.; Sharma, A. Super-resolution of Magnetic Resonance Images using deep Convolutional Neural Networks. In Proceedings of the IEEE International Conference on Consumer Electronics—Taiwan (ICCE-TW), Taipei, Taiwan, 12–14 June 2017; pp. 41–42.
24. Mesin, L.; Roatta, S.; Pasquero, P.; Porta, M. Automated Volume Status Assessment Using Inferior Vena Cava Pulsatility. *Electronics* **2020**, *9*, 1671. [CrossRef]
25. Mesin, L.; Pasquero, P.; Albani, S.; Porta, M.; Roatta, S. Semi-automated tracking and continuous monitoring of inferior vena cava diameter in simulated and experimental ultrasound imaging. *Ultrasound Med. Biol.* **2015**, *41*, 845–857. [CrossRef]
26. Mesin, L.; Pasquero, P.; Roatta, S. Tracking and Monitoring Pulsatility of a Portion of Inferior Vena Cava from Ultrasound Imaging in Long Axis. *Ultrasound Med. Biol.* **2019**, *45*, 1338–1343. [CrossRef] [PubMed]
27. Mesin, L.; Pasquero, P.; Roatta, S. Multi-directional assessment of Respiratory and Cardiac Pulsatility of the Inferior Vena Cava from Ultrasound Imaging in Short Axis. *Ultrasound Med. Biol.* **2020**, *46*, 3475–3482. [CrossRef] [PubMed]
28. Mesin, L.; Giovinazzo, T.; D'Alessandro, S.; Roatta, S.; Raviolo, A.; Chiacchiarini, F.; Porta, M.; Pasquero, P. Improved repeatability of the estimation of pulsatility of inferior vena cava. *Ultrasound Med. Biol.* **2019**, *45*, 2830–2843. [CrossRef] [PubMed]
29. Mesin, L.; Albani, S.; Sinagra, G. Non-invasive Estimation of Right Atrial Pressure using the Pulsatility of Inferior Vena Cava. *Ultrasound Med. Biol.* **2019**, *45*, 1331–1337. [CrossRef] [PubMed]
30. Albani, S.; Pinamonti, B.; Giovinazzo, T.; de Scordilli, M.; Fabris, E.; Stolfo, D.; Perkan, A.; Gregorio, C.; Barbati, G.; Geri, P.; et al. Accuracy of right atrial pressure estimation using a multi-parameter approach derived from inferior vena cava semi-automated edge-tracking echocardiography: A pilot study in patients with cardiovascular disorders. *Int. J. Cardiovasc. Imaging* **2020**, *36*, 1213–1225. [CrossRef]
31. Folino, A.; Benzo, M.; Pasquero, P.; Laguzzi, A.; Mesin, L. Messere, A.; Porta, M.; Roatta, S. Vena Cava Responsiveness to Controlled Isovolumetric Respiratory Efforts. *J. Ultrasound Med.* **2017**, *36*, 2113–2123. [CrossRef]

32. Chesbrough, H.W. *Open Innovation. The New Imperative for Creating and Profiting from Technology*; Harvard Business Review Press: Brighton, Boston, MA, USA, 2003.

Publisher's Note: MDPI stays neutral with regard to jurisdictional claims in published maps and institutional affiliations.

Article

Semantic Segmentation Framework for Glomeruli Detection and Classification in Kidney Histological Sections

Nicola Altini [1], Giacomo Donato Cascarano [1], Antonio Brunetti [1], Francescomaria Marino [1], Maria Teresa Rocchetti [2], Silvia Matino [2], Umberto Venere [2], Michele Rossini [2], Francesco Pesce [2], Loreto Gesualdo [2] and Vitoantonio Bevilacqua [1,*]

1 Department of Electrical and Information Engineering (DEI), Polytechnic University of Bari, 70126 Bari, Italy; nicola.altini@poliba.it (N.A.); giacomodonato.cascarano@poliba.it (G.D.C.); antonio.brunetti@poliba.it (A.B.); francescomaria.marino@poliba.it (F.M.)

2 Department of Emergency and Organ Transplantation (DETO), Nephrology Unit, University of Bari Aldo Moro, 70126 Bari, Italy; mariateresa.rocchetti@unifg.it (M.T.R.); silviamatino@gmail.com (S.M.); u.venere@gmail.com (U.V.); michelerossini@libero.it (M.R.); f.pesce81@gmail.com (F.P.); loreto.gesualdo@uniba.it (L.G.)

* Correspondence: vitoantonio.bevilacqua@poliba.it

Received: 28 February 2020; Accepted: 17 March 2020; Published: 19 March 2020

Abstract: the evaluation of kidney biopsies performed by expert pathologists is a crucial process for assessing if a kidney is eligible for transplantation. In this evaluation process, an important step consists of the quantification of global glomerulosclerosis, which is the ratio between sclerotic glomeruli and the overall number of glomeruli. Since there is a shortage of organs available for transplantation, a quick and accurate assessment of global glomerulosclerosis is essential for retaining the largest number of eligible kidneys. In the present paper, the authors introduce a Computer-Aided Diagnosis (CAD) system to assess global glomerulosclerosis. The proposed tool is based on Convolutional Neural Networks (CNNs). In particular, the authors considered approaches based on Semantic Segmentation networks, such as SegNet and DeepLab v3+. The dataset has been provided by the Department of Emergency and Organ Transplantations (DETO) of Bari University Hospital, and it is composed of 26 kidney biopsies coming from 19 donors. The dataset contains 2344 non-sclerotic glomeruli and 428 sclerotic glomeruli. The proposed model consents to achieve promising results in the task of automatically detecting and classifying glomeruli, thus easing the burden of pathologists. We get high performance both at pixel-level, achieving mean F-score higher than 0.81, and Weighted Intersection over Union (IoU) higher than 0.97 for both SegNet and Deeplab v3+ approaches, and at object detection level, achieving 0.924 as best F-score for non-sclerotic glomeruli and 0.730 as best F-score for sclerotic glomeruli.

Keywords: semantic segmentation; convolutional neural networks; kidney biopsy; kidney transplantation; glomerulus detection; glomerulosclerosis

1. Introduction

The spread of Deep Learning (DL) techniques and frameworks has led to a revolution in the medical imaging field. The assessment of organ viability, by donor kidney biopsy examination, is essential prior to transplantation. The traditional evaluation of biopsies was based on the visual analysis by trained pathologists of biopsy slides using a light microscope which is a time consuming and highly variable procedure. The high variability between the observers resulted in poor reproducibility among pathologists, which may cause an inappropriate organ discard.

Therefore, the development of new techniques able to objectively and rapidly interpret donor kidney biopsy to support pathologist's decision making is strongly fostered. The increasing availability of whole-slide scanners, which facilitate the digitization of histopathological tissue, led to a new research field denoted as digital pathology and generated a strong demand for the development of Computer-Aided Diagnosis (CAD) systems. As stated in the literature, the application of deep learning techniques for the analysis of Whole-Slide Images (WSIs) has shown significant results and suggest that the integration of DL framework with CAD systems is a valuable solution.

In the realm of digital pathology, several recent studies have proposed CAD systems for glomerulus identification and classification in renal biopsies [1–8]. The eligibility for transplantation of a kidney retrieved from Expanded Criteria Donors (ECD) relies on rush histological examination of the organ to evaluate suitability for transplant [9]. The Karpinski score is based on the microscopic examination of four compartments: glomerular, tubular, interstitial and vascular, in order to assess the degree of chronic injury. For each compartment is assigned a score from 0 to 3 where 0 corresponds to normal histology and 3 to the highest degree of, respectively, global glomerulosclerosis, tubular atrophy, interstitial fibrosis and arterial and arteriolar narrowing [9,10]. The evaluation of global glomerulosclerosis requires detection and classification of all the glomeruli present in a kidney biopsy, distinguishing between healthy (non-sclerotic) and non-healthy (sclerotic) ones.

The two fundamental components that characterize a non-sclerotic glomerulus are the capillary tuft with the mesangium and the Bowman's capsule. The first one is placed inside the glomerulus while the second one is peripheral and has the function to contain the tuft. The space between these two components is called Bowman's space. From a morphological point of view, the non-sclerotic glomerulus generally has an elliptic form. The capillary tuft has a pomegranate form, caused by the contemporary presence of blue points (nuclei of cells), white areas (capillary lumens) and variable amount of regions with similar tonality and different levels of saturation (mesangial matrix). A non-healthy glomerulus, from the point of view of Karpinski's score, is a globally sclerotic glomerulus, namely a glomerulus where capillary lumens are completely obliterated for increase in extracellular matrix and Bowman's space is completely filled by collagenous material. Examples of non-sclerotic and sclerotic glomeruli are depicted in Figure 1.

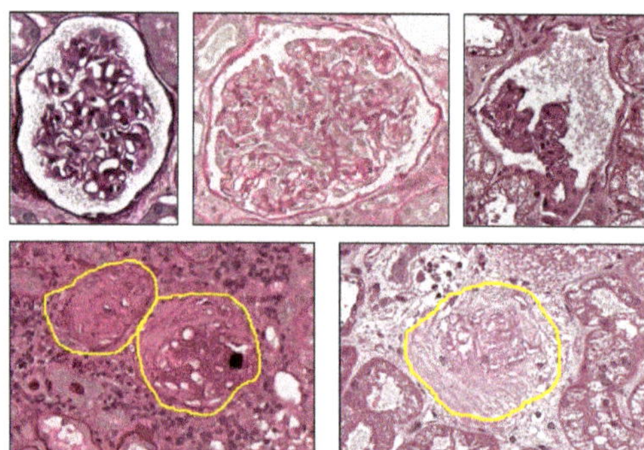

Figure 1. Glomeruli. Top row: non-sclerotic glomeruli. Bottom row: sclerotic glomeruli.

Ledbetter et al. proposed a Convolutional Neural Network to predict kidney function (evaluated as the quantity of primary filtrate that passes from the blood through the glomeruli per minute) in chronic kidney disease patients from whole-slide images of their kidney biopsies [3]. Gallego et al. proposed a method based on the pretrained AlexNet model [11] to perform glomerulus classification and detection in kidney tissue segments [2]. Gadermayr et al. focused on the segmentation of the glomeruli. The authors proposed two different CNN cascades for segmentation applications with sparse objects. They applied these approaches to the glomerulus segmentation task and compared them with conventional fully convolutional networks, coming to the conclusion that cascade networks can be a powerful tool for segmenting renal glomeruli [4]. Temerinac-Ott et al. compared the performance between a CNN classifier and a support-vector machines (SVM) classifier which exploits features extracted by histogram of oriented gradients (HOG) [12] for the task of glomeruli detection in WSIs with multiple stains, using a sliding window approach. The obtained results showed that the CNN method outperformed the HOG and SVM classifier [1]. Kawazoe et al. faced the task of glomeruli detection in multistained human kidney biopsy slides by using a Deep Learning approach based on Faster R-CNN [6]. Marsh et al. developed a deep learning model that recognizes and classifies sclerotic and non-sclerotic glomeruli in whole-slide images of frozen donor kidney biopsies. They used a Fully Convolutional Network (FCN) followed by a blob-detection algorithm [13], based on Laplacian-of-Gaussian, to post-process the FCN probability maps into object detection predictions [8]. Ginley et al. proposed a CAD to classify renal biopsies of patients with diabetic nephropathy [7], using a combination of classical image processing and novel machine learning techniques. Hermsen et al. adopted CNNs, namely an ensemble of five U-Nets, for segmentation of ten tissue classes from WSIs of periodic acid-Schiff (PAS) stained kidney transplant biopsies [14].

The analysis of the literature suggests that main works focused on the glomerular detection task only, without considering the further classification into sclerotic and non-sclerotic [1,2,4,6]. Few papers considered the assessment of global glomerulosclerosis from kidney biopsies [7,8,14].

In our previous works we focused on other kidney biopsies analysis tasks, such as classification of tubules and vessels [15] and classification of non-sclerotic and sclerotic glomeruli [5]. In this work, we propose a CAD system to address the segmentation and the classification tasks of glomeruli, in order to obtain a reliable estimate of Karpinski histological score. The proposed work allowed us to obtain better results than the literature in the classification task.

2. Materials

The kidney biopsies dataset analyzed in this paper has been provided by the Department of Emergency and Organ Transplantations (DETO) of the Bari University Hospital. Slides were digitized using a high-resolution whole-slide scanner with a scanning objective which has a 20× magnification corresponding to a resolution of 0.50 µm/pixel. All the biopsies provided by DETO clinicians are PAS stained sections from formalin fixed paraffin embedded tissue. The complete dataset is composed of 26 kidney biopsies coming from 19 donors. It contains 2344 non-sclerotic glomeruli and 428 sclerotic glomeruli. The dataset has been split into a train-validation (trainval) set and a test set. The trainval set has been further split into a train set and a validation set; the last one is used for tuning hyperparameters and for assessing the trend of the loss function and of accuracy during the training process. A detailed overview of the dataset is reported in Table 1.

Table 1. Dataset info.

Set	WSIs	Non-Sclerotic	Sclerotic	Ratio
Trainval set	19	1852	341	5.43 : 1
Test set	7	492	87	5.66 : 1
Dataset	26	2344	428	5.48 : 1

3. Methods

3.1. Semantic Segmentation Framework

Convolutional Neural Networks have had a widespread adoption in all kinds of image analysis tasks, starting from AlexNet which won ImageNet Large Scale Visual Recognition Challenge 2012 (ILSVRC 2012) [16] by a huge margin [11], though pioneering work was already done by LeCun much earlier for handwritten digit recognition [17].

Semantic segmentation is a task which consists of classifying all the pixels belonging to an input image. In order to accomplish this task, most CNN semantic segmentation architectures are based on encoder-decoder networks. The encoder is devoted to the feature extraction process, shrinking the spatial dimensions while increasing the depth. The decoder has the task to recover the spatial information from the output of the encoder. Due to the several application in the medical imaging field, in this work we considered two main approaches based on SegNet and DeepLab v3+ architectures. The main SegNet applications regard segmentation tasks such as semantic segmentation of prostate cancer [18], gland segmentation from colon cancer histology images [19] and brain tumor segmentation from multi-modal magnetic resonance images [20]. DeepLab v3+ has been used for the semantic segmentation of colorectal polyps [21] and the automatic liver segmentation [22,23].

SegNet is a CNN architecture for semantic segmentation proposed by researchers at University of Cambridge [24]. As other semantic segmentation architectures, SegNet is composed of an encoder network and a corresponding decoder network, followed by a final pixel-wise classification layer. One clever point of SegNet is that it removes the necessity of learning the upsampling process, by storing indices used in max-pooling step in encoder and applying them when upsampling in the corresponding layers of the decoder.

DeepLab is an architecture proposed by Chen et al. [25]. One of the interesting novelties proposed by the authors of DeepLab is the atrous convolution, also known as dilated convolution. The idea has been commonly used in wavelet transform before being adapted to convolutions for deep learning. Atrous convolution consents to broaden the field of view of filters to incorporate larger context. It is, therefore, a valuable tool to tune the field of view, permitting identification of the right balance between context assimilation (large field of view) and fine localization (small field of view). We adopted DeepLab v3+ [26] with ResNet-18 [27] as backbone in our tests.

We replaced the last layer of both SegNet and DeepLab v3+ networks with a pixel-wise classification layer with 3 output classes (background, sclerotic glomeruli and non-sclerotic glomeruli); we used inverse class frequencies as class weights and pixel-wise cross-entropy as loss function.

3.2. *Proposed Workflow*

3.2.1. CAD Architecture

A high-level overview of the proposed CAD is depicted in Figure 2. The physicians can visualize the WSIs using Aperio ImageScope software. In order to perform supervised learning, we need labelled data. Pathologists can annotate the slides using ImageScope, and export the results in XML files, which we can use to feed our neural networks. After having trained our models, we can export the output in XML files, and physicians can see the CAD annotations always in ImageScope, with seamless integration. To accomplish the task of calculating the Karpinski histological score, we must make a careful choice for the architecture of the network. All the models have been trained and validated on a machine with the characteristics reported in Table 2.

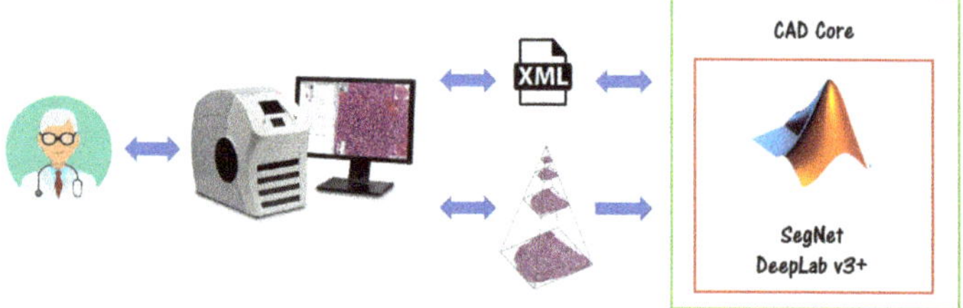

Figure 2. CAD architecture. Physicians can visualize and annotate the WSIs using Aperio ImageScope software. The developed Deep Learning models can interact with ImageScope through an XML interface.

Table 2. System Details.

	System Details
GPU	NVIDIA GTX 1060 with 6 GB of RAM
CPU	Intel Core i7-4790 CPU @ 3.60 Ghz
RAM	32 GB
OS	Microsoft Windows 10 Home
Tool	MATLAB R2019a

3.2.2. Semantic Segmentation Workflow

To obtain an estimate of the Karpinski score, we must detect and classify all the glomeruli which appear in the WSI. We first use a semantic segmentation CNN to obtain a pixel-level classification, distinguishing between pixels which belongs to background, sclerotic and non-sclerotic glomeruli. Then, we must turn these pixel-level classifications into object detections, so that we can count the number of sclerotic and non-sclerotic glomeruli. The general schema for our semantic segmentation-based glomerular detector is depicted in Figure 3.

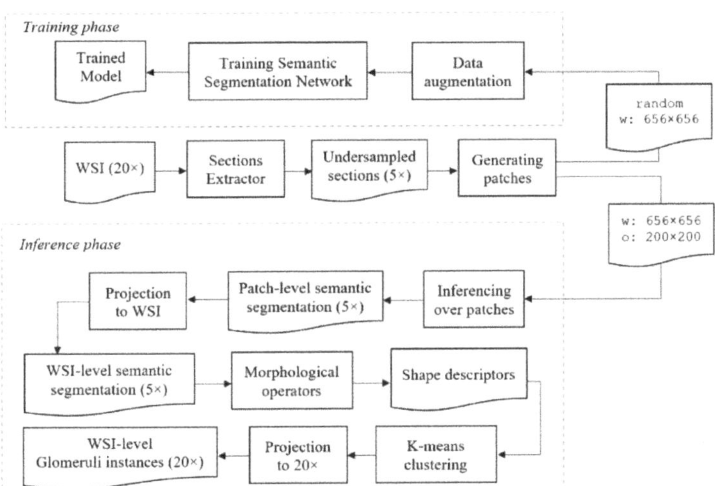

Figure 3. Semantic Segmentation approach architecture. The top part describes how to train the CNN. The bottom part explains how to use the trained model for performing inference, and the related morphological and clustering post-processing steps.

The first step in our workflow consists of segmenting the sections present in the WSI. At this purpose, we used classical Image Processing techniques as thresholding, morphological operators, connected components labelling, and eventually, clustering. A similar preprocessing step has also been done by Ledbetter et al. [3]. We refer to the module performing this step as Sections Extractor. To reduce the very large dimension of WSIs, which can be overwhelming for Deep Learning algorithms, we undersampled the sections by a factor of 4. The original WSIs have a magnification of 20×, after undersampling it becomes equivalent to a magnification of 5×. This operation leads to an effective downsampling of the images from a resolution of about 8000 × 8000 pixels to a resolution of about 2000 × 2000 pixels. Since the section obtained this way was still too large to fit in our GPU, we divided it in patches. During training, we randomly sampled patches of size 656 × 656, with a mechanism to avoid to take too many patches only with negatives samples. The random patches sampled during the training process are then fed to a data augmentation block that performs different augmentations, as reported in Table 3. Augmentations are generated on-the-fly for each epoch within random ranges, so the network always processes slightly different input data, thus reducing the risk of overfitting. In the inference phase, we take patches of size 656 × 656 pixels, with an overlap between successive windows of 200 × 200 pixels. Please note that in semantic segmentation is important to have a larger context for performing inference, when the approach involves a sliding window processing [28]. After we get the predicted masks for glomeruli at patch-level, we project them to the original WSI, to get the WSI-level predicted mask. At this point, we apply morphological operators to remove noisy points and smooth the glomeruli shapes. We then analyze shape descriptors to understand if it is necessary to perform a clustering operation. In the end, the obtained mask is projected to 20× resolution, corresponding to oversampling by 4, using nearest-neighbour interpolation. Please note that in this work, all the resizing operations involving the digital pathology images are obtained using bicubic interpolation, while all the resizing operations involving the categorical masks are obtained using nearest-neighbour interpolation.

3.2.3. Morphological Operators and Clustering

Adapting a semantic segmentation network to perform object detection poses some challenges. The task of semantic segmentation consists of labelling only individual pixels, which mainly captures textural information. In contrast to architectures explicitly tailored to Object Detection, such as Faster R-CNN [29] or Mask R-CNN [30], where there are anchor boxes, the network does not look for objects, it just tries to classify individual pixels. To extend the semantic segmentation model into an instance segmentation one, we must use different morphological operators and clustering algorithms as post-processing steps.

Morphological operators are applied only to binary masks obtained as the output of the semantic segmentation networks. First, we smooth the shapes of objects performing a morphological closing operation, with a disk of radius 5 pixels as structuring element, and with the morphological flood-fill operation. Then, we delete small objects and noisy points using opening operator, with a disk of radius 10 pixels as structuring element, and area opening operator, removing connected regions with an area below 1000 pixels. Examples are depicted in Figure 4, where binary masks are overlapped to the biopsy images for visualization purposes. Masks relative to non-sclerotic and sclerotic glomeruli are green and red colored, respectively. Lastly, we analyze the shape descriptors for each of these objects to understand if there are touching objects we need to cluster. The sequence of morphological operators used is depicted in Figure 5.

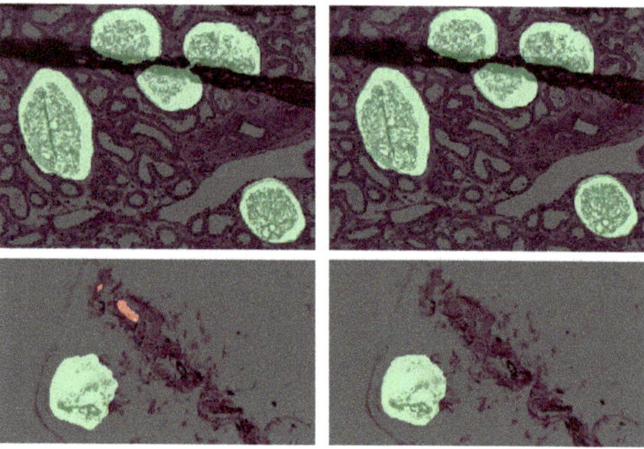

Figure 4. (**Left**) Semantic Segmentation output. (**Right**) After Morphological Operators.

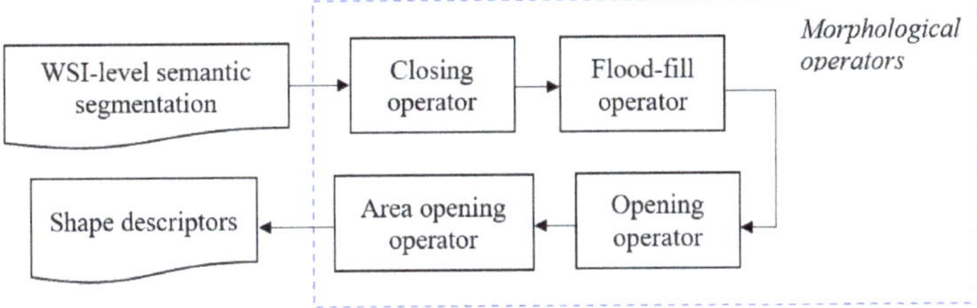

Figure 5. Morphological operators sequence applied to the output masks from the semantic segmentation network. The output of the morphological post-processing is used for calculating shape descriptors to eventually perform clustering.

An important observation is that individual glomeruli have convex shapes, so their area is pretty similar to their convex area. We perform a *K*-means clustering based on the difference between the convex area and the area, as specified in Equation (1).

$$deltaArea = convexHullArea - area \tag{1}$$

We decide the number *K* of clusters according to `deltaArea`: if `deltaArea` $\leq$ 900, $K = 1$; if `deltaArea` $>$ 900 and `deltaArea` $\leq$ 5000, $K = 2$; if `deltaArea` $>$ 5000, $K = 3$. The values of `deltaArea` and the corresponding *K* have been empirically determined on the trainval set. Confusion matrices reported later have been obtained after the clustering with the configuration based on `deltaArea`. Examples of glomeruli before clustering are depicted in Figure 6a,c. The corresponding images after clustering are shown in Figure 6b,d.

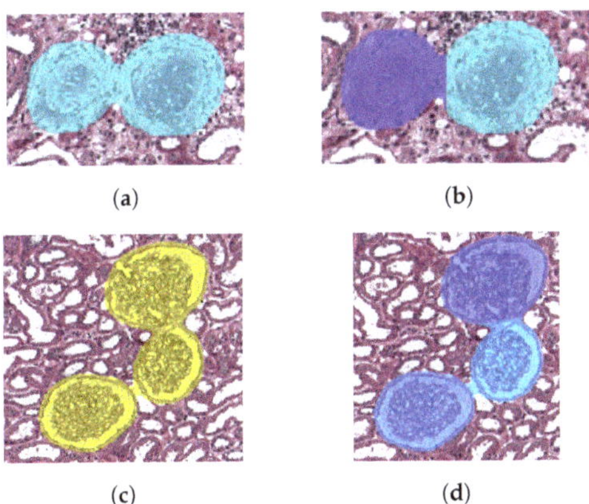

Figure 6. Examples of K-means clustering for both sclerotic and non-sclerotic glomeruli. The number *K* of clusters is determined according to `deltaArea` defined in (1). **(a)** Sclerotic glomeruli before clustering. **(b)** Sclerotic glomeruli after clustering, with $K = 2$. **(c)** Non-sclerotic glomeruli before clustering. **(d)** Non-sclerotic glomeruli after clustering, with $K = 3$.

3.2.4. Data Augmentation

Tellez et al. analyzed the problem of stain color variation in digital pathology very deeply [31]. They proposed different solutions for both stain color augmentation and stain color normalization. In this work, we exploited techniques proposed by them such as morphological transformations and Hue-Saturation-Value (HSV) shifts. An interesting morphological transformation is the elastic deformation; it was originally proposed by Simard et al. [32] for the analysis of visual documents, and then has had a widespread application in medical imaging, as also shown by U-Net authors [28]. We used elastic deformation to generate plausible alterations of glomeruli shapes, increasing the variability of training images and thus reducing the risk of overfitting. An example of elastic deformation applied to our images is depicted in Figure 7. Examples of HSV shift are depicted in Figure 8.

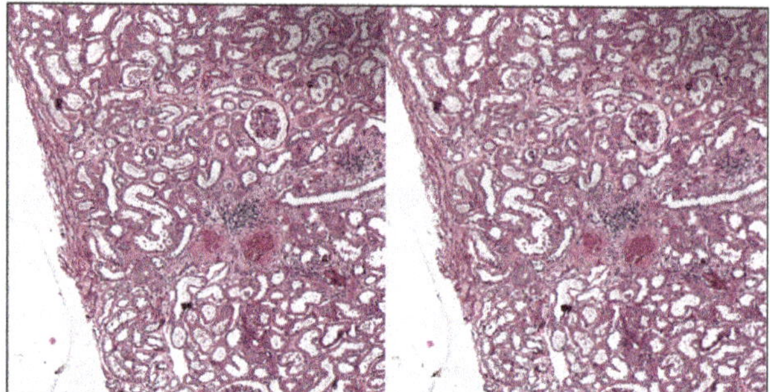

Figure 7. Elastic deformation example. Left: original image. Right: after elastic deformation with $\sigma = 6.29$, $\alpha = 340$.

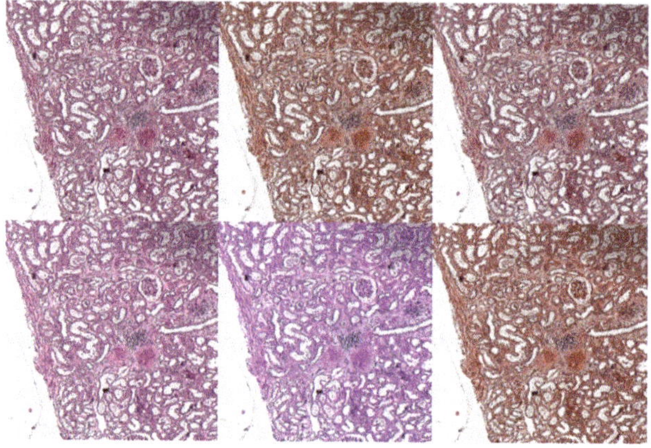

Figure 8. HSV shift examples. Top Left: original image. Top Center: $\Delta H = +0.18$, $\Delta S = +0.03$. Top Right: $\Delta H = +0.06$, $\Delta S = -0.06$. Bottom Left: $\Delta H = -0.04$, $\Delta S = -0.02$. Bottom Center: $\Delta H = -0.11$, $\Delta S = +0.10$. Bottom Right: $\Delta H = +0.18$, $\Delta S = +0.09$.

A summary of the data augmentation techniques used for the training process is reported in Table 3. The augmentations in group 1 are independently performed, each with a given probability p. Resize augmentation used here is slightly different from standard resize; in fact, we apply mirroring padding (instead of zero padding) when we perform a resize which shrinks the image size. Augmentations, such as mirroring padding, which alter the morphology of the image are also executed for the mask. From the augmentations reported in group 2, only one is made. Group 3 contains only one augmentation, which is performed with a given probability. The augmentations are performed in the order they compare in the table, i.e., before the four in group 1, then one of group 2 and in the end the one of group 3.

Table 3. Augmentations.

Data Augmentation	
Type	Details
Group 1	
Rotate	$\theta = 90$, $p = 0.25$
Flip left-right	$p = 0.25$
Flip upside-down	$p = 0.25$
Resize	$resize \in [0.8, 1.2]$, $p = 0.25$
Group 2	
Gaussian Noise	$\sigma \in [0, 0.01]$, $p = 0.1$
Gaussian Blur	$\sigma \in [0, 0.1]$, $p = 0.1$
Elastic Deformation	$\sigma \in [2, 5]$, $\alpha \in [100, 300]$, $p = 0.2$
Group 3	
HSV shift	$\Delta S \in [-0.1, 0.1]$, $\Delta H \in [-0.1, 0.1]$, $p = 0.5$

3.2.5. Hyperparameters Tuning

We tried different semantic segmentation network architectures. For SegNet and DeepLab v3+ we tuned hyperparameters according to Table 4. Please note that DeepLab v3+ with ResNet-18 backbone is more lightweight than SegNet, and this allowed us to use a larger mini-batch size, with eight patches per mini-batch. With our GPU, SegNet was trained with only one patch per mini-batch. More details about hyperparameters can be found in MATLAB documentation [33].

Table 4. Hyperparameters.

Hyperparameter	**SegNet**	**Deeplab v3+**
Optimizer	SGDM	SGDM
LearnRateSchedule	'piecewise'	'piecewise'
LearnRateDropPeriod	10	10
LearnRateDropFactor	0.3	0.3
Momentum	0.9	0.9
InitialLearnRate	0.001	0.001
L2Regularization	0.005	0.005
MaxEpochs	30	30
MiniBatchSize	1	8
Shuffle	'every-epoch'	'every-epoch'
ValidationPatience	10	10
ValidationFrequency	1 per epoch	1 per epoch

4. Experimental Results

We distinguish between the results obtained at pixel-level (semantic segmentation task) and at object detection level.

In particular, for the semantic segmentation task we group the metrics in Dataset Metrics and Class Metrics [33].

The group of **Dataset Metrics** includes semantic segmentation metrics aggregated over the data set: *Global Accuracy*, *Mean Accuracy* (the mean of the accuracies calculated per class), *Mean IoU* (the mean of the IoUs calculated per class), *Weighted IoU* (mean of the IoUs, weighted by the number of pixels in the class) and *Mean F-score* (mean of the F-measures calculated per class).

The group of **Class Metrics** includes semantic segmentation metrics calculated for each class, namely: *Accuracy* (2), *IoU* (3) and *Mean F-score* (F-measure for each class, averaged over all images).

$$Accuracy = \frac{TP + TN}{TP + TN + FP + FN} \tag{2}$$

$$IoU = \frac{TP}{TP + FP + FN} \tag{3}$$

For the object detection task, confusion matrices are calculated assuming that a true positive match between predicted mask and ground truth mask has pixel-wise IoU (3) of at least 0.2. Besides confusion matrices, the metrics used for assessing the results of the object detection task are:

$$Precision = \frac{TP}{TP + FP}, \tag{4}$$

$$Recall = \frac{TP}{TP + FN}, \tag{5}$$

$$F_1\ Score = \frac{2 \cdot Precision \cdot Recall}{Precision + Recall}. \tag{6}$$

The best results on non-sclerotic glomeruli have been obtained using DeepLab v3+, while for sclerotic glomeruli the best model was SegNet. An example of the output of our semantic segmentation framework is depicted in Figure 9.

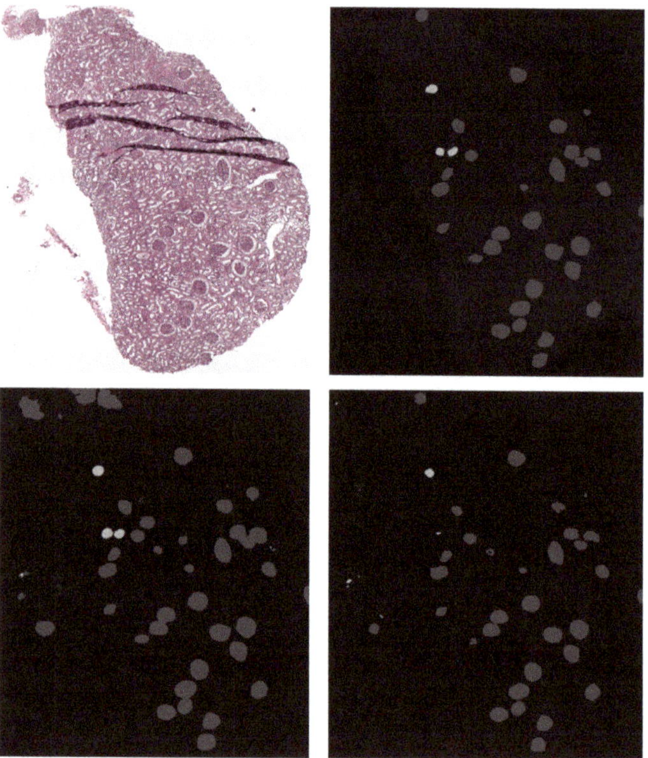

Figure 9. Top Left: original image. Top Right: ground truth. Bottom Left: SegNet prediction. Bottom Right: DeepLab v3+ prediction. Sclerotic glomeruli and non-sclerotic ones are white and gray colored, respectively.

4.1. Pixel-Level Metrics

Pixel-level dataset metrics for both SegNet and DeepLab v3+ are reported in Table 5. The pixel-level class metrics of SegNet and DeepLab v3+ are reported in Tables 6 and 7, respectively.

The normalized pixel-level confusion matrix are in Tables 8 and 9. Pixel-level confusion matrices are normalized per row; B, NS, S stand for Background, Non-sclerotic and sclerotic, respectively.

Table 5. Dataset Metrics.

CNN	Global Accuracy	Mean Accuracy	Mean IoU	Weighted IoU	Mean F-Score
SegNet	0.98346	0.86385	0.71352	0.97156	0.81784
Deeplab v3+	0.99179	0.76884	0.72873	0.98434	0.84614

Table 6. Class Metrics SegNet.

Class	Accuracy	IoU	Mean F-Score
Background	0.98636	0.98294	0.99243
Non-sclerotic	0.91925	0.66546	0.83239
sclerotic	0.68594	0.49215	0.69686

Table 7. Class Metrics Deeplab v3+.

Class	Accuracy	IoU	Mean F-Score
Background	0.99690	0.99172	0.96684
Non-sclerotic	0.88199	0.80872	0.93306
Sclerotic	0.42764	0.38574	0.63852

Table 8. Normalized pixel-level Confusion Matrix SegNet.

		Prediction		
		B	NS	S
Ground Truth	B	98.64%	1.26%	0.10%
	NS	8.07%	91.93%	0.00%
	S	30.97%	0.44%	68.59%

Table 9. Normalized pixel-level Confusion Matrix Deeplab v3+.

		Prediction		
		B	NS	S
Ground Truth	B	99.69%	0.28%	0.03%
	NS	11.78%	88.20%	0.02%
	S	50.57%	6.67%	42.76%

4.2. Object Detection Metrics

In object detection confusion matrices B, NS, S stand for Background, Non-sclerotic and Sclerotic, respectively.

The object detection confusion matrices for SegNet and DeepLab v3+ are reported in Tables 10 and 11, respectively. The detection metrics for both the proposed models and a comparison with the method proposed by Marsh et al. [8] are reported in Table 12. The SegNet-based model obtained a better F-score for both the glomeruli classes. The DeepLab v3+-based model obtained a better F-score for non-sclerotic glomeruli and a slightly worse F-score for sclerotic glomeruli.

Table 10. Object Detection Confusion Matrix SegNet.

		Prediction		
		NS	S	B
Ground Truth	NS	436	0	56
	S	1	58	28
	B	86	14	-

Table 11. Object Detection Confusion Matrix Deeplab v3+.

		Prediction		
		NS	S	B
Ground Truth	NS	449	0	43
	S	7	41	39
	B	24	1	-

Table 12. Performance Comparison for Detection Metrics.

Author	Model	Class	Recall	Precision	F-Score
Marsh et al. [8]	FCN + blob-detection	NS	0.885	0.813	0.848
		S	0.698	0.607	0.649
Proposed approach	SegNet	NS	0.886	0.834	0.859
		S	0.667	0.806	0.730
	DeepLab v3+	NS	0.913	0.935	0.924
		S	0.471	0.976	0.636

5. Conclusions and Future Work

The proposed approach allowed us to obtain high performance both at pixel and object detection level. The semantic segmentation achieved mean F-score higher than 0.81 and Weighted IoU higher than 0.97 for both SegNet and Deeplab v3+ approaches; the glomeruli detection achieved 0.924 as best F-score for non-sclerotic glomeruli and 0.730 as best F-score for sclerotic glomeruli. We compared our obtained performance with the state of the art. As stated in the Section 1, there are three main works that face the problem of glomerular classification. Ginley et al. considered the glomerular assessment for patients affected by diabetic nephropathy but not for transplantation purposes [7]. Hermsen et al. considered many tissue classes, but the number of sclerotic glomeruli in their datasets is too small for a comparison with our method [14]. Marsh et al. considered the problem of global glomerulosclerosis from kidney transplant biopsies with haematoxylin and eosin (HE) stain [8]. The performance comparison between our proposed methods and Marsh et al. work is reported in Table 12. The obtained results show an improvement over the work of Marsh et al. Thus, CNNs for Semantic Segmentation are a viable approach for the purpose of glomerular segmentation and classification, allowing the obtaining of a reliable estimate of the global glomerulosclerosis. Assessing the suitability of kidney from ECD donors relies in many centers on the histological examination of kidney biopsies performed at the time of organ retrieval and processed and evaluated by on-call pathologist that, not necessarily, is an expert trained in renal pathology. The importance of training in renal pathology when assessing biopsy of such cases has been evaluated in some studies reporting better correlation with subsequent allograft outcome of histological scores provided by renal pathologists compared to those provided by general pathologist with potential risk of "overscoring" and the potential of discarding kidneys that could have been potentially transplanted [34–36]. The results were validated by the renal pathologists which assessed the reliability of the proposed workflow; the applied methodology constitutes a milestone in the creation of a CAD system for the renal transplant assessment. The proposed system could help pathologists in accomplishing the laborious task of evaluating the eligibility of a kidney for transplantation, providing a rapid and accurate result. Future work will include the use of Deep Learning models explicitly designed for the detection task, such as Faster R-CNN and Mask R-CNN.

Author Contributions: Conceptualization, N.A., G.D.C. and V.B.; Data curation, M.R., F.P. and L.G.; Methodology, N.A. and G.D.C.; Supervision, F.M., L.G. and V.B.; Validation, M.T.R., M.R., F.P. and L.G.; Writing—original draft, N.A.; Writing—review & editing, G.D.C., A.B., F.M., M.T.R., S.M., U.V., M.R., F.P., L.G. and V.B. All authors have read and agreed to the published version of the manuscript.

Funding: This work has been partially funded by the Italian Apulian Region project "SOS – Smart Operating Shelter" (INNONETWORK n. 9757YR7).

Conflicts of Interest: The authors declare no conflict of interest.

References

1. Temerinac-Ott, M.; Forestier, G.; Schmitz, J.; Hermsen, M.; Bräsen, J.; Feuerhake, F.; Wemmert, C. Detection of glomeruli in renal pathology by mutual comparison of multiple staining modalities. In Proceedings of the IEEE 10th International Symposium on Image and Signal Processing and Analysis, Ljubljana, Slovenia, 18–20 Sepember 2017; pp. 19–24.
2. Gallego, J.; Pedraza, A.; Lopez, S.; Steiner, G.; Gonzalez, L.; Laurinavicius, A.; Bueno, G. Glomerulus classification and detection based on convolutional neural networks. *J. Imaging* **2018**, *4*, 20. [CrossRef]
3. Ledbetter, D.; Ho, L.; Lemley, K.V. *Prediction of Kidney Function from Biopsy Images Using Convolutional Neural Networks*; Los Alamos National Lab: Santa Fe, NM, USA, 2017; pp. 1–11.
4. Gadermayr, M.; Dombrowski, A.K.; Klinkhammer, B.M.; Boor, P.; Merhof, D. CNN Cascades for Segmenting Whole Slide Images of the Kidney. *arXiv* **2017**, arXiv:1708.00251.
5. Cascarano, G.D.; Debitonto, F.S.; Lemma, R.; Brunetti, A.; Buongiorno, D.; De Feudis, I.; Guerriero, A.; Rossini, M.; Pesce, F.; Gesualdo, L.; et al. *An Innovative Neural Network Framework for Glomerulus Classification Based on Morphological and Texture Features Evaluated in Histological Images of Kidney Biopsy*; Springer: Berlin, Germany, 2019; pp. 727–738._66. [CrossRef]
6. Kawazoe, Y.; Shimamoto, K.; Yamaguchi, R.; Shintani-Domoto, Y.; Uozaki, H.; Fukayama, M.; Ohe, K. Faster R-CNN-based glomerular detection in multistained human whole slide images. *J. Imaging* **2018**, *4*, 91. [CrossRef]
7. Ginley, B.; Lutnick, B.; Jen, K.Y.; Fogo, A.B.; Jain, S.; Rosenberg, A.; Walavalkar, V.; Wilding, G.; Tomaszewski, J.E.; Yacoub, R.; et al. Computational Segmentation and Classification of Diabetic Glomerulosclerosis. *J. Am. Soc. Nephrol.* **2019**, *30*, 1953–1967. [CrossRef] [PubMed]
8. Marsh, J.N.; Matlock, M.K.; Kudose, S.; Liu, T.C.; Stappenbeck, T.S.; Gaut, J.P.; Swamidass, S.J. Deep learning global glomerulosclerosis in transplant kidney frozen sections. *IEEE Trans. Med. Imaging* **2018**, *37*, 2718–2728. [CrossRef] [PubMed]
9. Karpinski, J.; Lajoie, G.; Cattran, D.; Fenton, S.; Zaltzman, J.; Cardella, C.; Cole, E. Outcome of kidney transplantation from high-risk donors is determined by both structure and function. *Transplantation* **1999**, *67*, 1162–1167. [CrossRef]
10. Remuzzi, G.; Grinyò, J.; Ruggenenti, P.; Beatini, M.; Cole, E.H.; Milford, E.L.; Brenner, B.M. Early experience with dual kidney transplantation in adults using expanded donor criteria. *J. Am. Soc. Nephrol.* **1999**, *10*, 2591–2598.
11. Krizhevsky, A.; Sutskever, I.; Hinton, G.E. 2012 AlexNet. In *Advances in Neural Information Processing Systems*; MIT Press: Cambridge, MA, USA, 2012. [CrossRef]
12. Dalal, N.; Triggs, B. Histograms of oriented gradients for human detection. In Proceedings of the IIEEE Computer Society Conference on Computer Vision and Pattern Recognition (CVPR'05), San Diego, CA, USA, 20–26 June 2005; Volume 1, pp. 886–893.
13. Lindeberg, T. Detecting salient blob-like image structures and their scales with a scale-space primal sketch: A method for focus-of-attention. *Int. J. Comput. Vis.* **1993**, *11*, 283–318. [CrossRef]

14. Hermsen, M.; de Bel, T.; Den Boer, M.; Steenbergen, E.J.; Kers, J.; Florquin, S.; Roelofs, J.J.; Stegall, M.D.; Alexander, M.P.; Smith, B.H.; et al. Deep Learning–Based Histopathologic Assessment of Kidney Tissue. *J. Am. Soc. Nephrol.* **2019**, *30*, 1968–1979. [CrossRef]
15. Bevilacqua, V.; Pietroleonardo, N.; Triggiani, V.; Brunetti, A.; Di Palma, A.M.; Rossini, M.; Gesualdo, L. An innovative neural network framework to classify blood vessels and tubules based on Haralick features evaluated in histological images of kidney biopsy. *Neurocomputing* **2017**, *228*, 143–153. [CrossRef]
16. Russakovsky, O.; Deng, J.; Su, H.; Krause, J.; Satheesh, S.; Ma, S.; Huang, Z.; Karpathy, A.; Khosla, A.; Bernstein, M.; et al. Imagenet large scale visual recognition challenge. *Int. J. Comput. Vis.* **2015**, *115*, 211–252. [CrossRef]
17. Lecun, Y.; Bottou, L.; Bengio, Y.; Ha, P. LeNet. *Proc. IEEE* **1998**. [CrossRef]
18. Ma, Z.; Li, J.; Salemi, H.; Arnold, C.; Knudsen, B.S.; Gertych, A.; Ing, N. Semantic segmentation for prostate cancer grading by convolutional neural networks. In Proceedings of the SPIE Medical Imaging 2018: Digital Pathology, Houston, TX, USA, 10–15 February 2018; p. 46. [CrossRef]
19. Tang, J.; Li, J.; Xu, X. Segnet-based gland segmentation from colon cancer histology images. In Proceedings of the IEEE 33rd Youth Academic Annual Conference of Chinese Association of Automation (YAC), Jiangsu, China, 18–20 May 2018; pp. 1078–1082.
20. Alqazzaz, S.; Sun, X.; Yang, X.; Nokes, L. Automated brain tumor segmentation on multi-modal MR image using SegNet. *Comput. Visual Media* **2019**, *5*, 209–219. [CrossRef]
21. Xiao, W.T.; Chang, L.J.; Liu, W.M. Semantic segmentation of colorectal polyps with DeepLab and LSTM networks. In Proceedings of the IEEE International Conference on Consumer Electronics-Taiwan (ICCE-TW), Taiwan, China, 19–21 May 2018; pp. 1–2.
22. Tang, W.; Zou, D.; Yang, S.; Shi, J. DSL: Automatic liver segmentation with faster R-CNN and DeepLab. In Proceedings of the International Conference on Artificial Neural Networks, Rhodes, Greece, 4–7 October 2018; Springer: Berlin, Germany, 2018; pp. 137–147.
23. Tang, W.; Zou, D.; Yang, S.; Shi, J.; Dan, J.; Song, G. A two-stage approach for automatic liver segmentation with Faster R-CNN and DeepLab. *Neural Comput. Appl.* **2020**, 1–10. [CrossRef]
24. Badrinarayanan, V.; Kendall, A.; Cipolla, R. SegNet: A Deep Convolutional Encoder-Decoder Architecture for Image Segmentation. *IEEE Trans. Pattern Anal. Mach. Intell.* **2017**, *39*, 2481–2495. [CrossRef] [PubMed]
25. Chen, L.C.; Papandreou, G.; Kokkinos, I.; Murphy, K.; Yuille, A.L. DeepLab: Semantic Image Segmentation with Deep Convolutional Nets, Atrous Convolution, and Fully Connected CRFs. *IEEE Trans. Pattern Anal. Mach. Intell.* **2018**, *40*, 834–848. [CrossRef]
26. Chen, L.C.; Zhu, Y.; Papandreou, G.; Schroff, F.; Adam, H. Encoder-decoder with atrous separable convolution for semantic image segmentation. In *Lecture Notes in Computer Science (including subseries Lecture Notes in Artificial Intelligence and Lecture Notes in Bioinformatics)*; Springer: Berlin, Germany, 2018; Volume 11211 LNCS, pp. 833–851._49. [CrossRef]
27. He, K.; Zhang, X.; Ren, S.; Sun, J. Deep residual learning for image recognition. In Proceedings of the IEEE Computer Society Conference on Computer Vision and Pattern Recognition, Las Vegas, NV, USA, 27–30 June 2016; pp. 770–778. [CrossRef]
28. Ronneberger, O.; Fischer, P.; Brox, T. U-net: Convolutional networks for biomedical image segmentation. In *Lecture Notes in Computer Science (Including Subseries Lecture Notes in Artificial Intelligence and Lecture Notes in Bioinformatics)*; Springer: Berlin, Germany, 2015; Volume 9351, pp. 234–241._28. [CrossRef]
29. Ren, S.; He, K.; Girshick, R.; Sun, J. Faster R-CNN: Towards Real-Time Object Detection with Region Proposal Networks. *IEEE Trans. Pattern Anal. Mach. Intell.* **2017**, *39*, 1137–1149. [CrossRef]
30. He, K.; Gkioxari, G.; Dollar, P.; Girshick, R. Mask R-CNN. In Proceedings of the IEEE International Conference on Computer Vision, Rio, Brazil, 14–21 October 2017. [CrossRef]
31. Tellez, D.; Litjens, G.; Bándi, P.; Bulten, W.; Bokhorst, J.M.; Ciompi, F.; van der Laak, J. Quantifying the effects of data augmentation and stain color normalization in convolutional neural networks for computational pathology. *Med. Image Anal.* **2019**, *58*, 101544. [CrossRef]
32. Simard, P.Y.; Steinkraus, D.; Platt, J.C. Best Practices for Convolutional Neural Networks Applied to Visual Document Analysis. In Proceedings of the ICDAR, Edinburgh, UK, 3–6 August 2003; Volume 3.
33. MATLAB Documentation. Available online: https://www.mathworks.com/help/matlab/ (accessed on 19 November 2019).

34. Azancot, M.A.; Moreso, F.; Salcedo, M.; Cantarell, C.; Perello, M.; Torres, I.B.; Montero, A.; Trilla, E.; Sellarés, J.; Morote, J.; et al. The reproducibility and predictive value on outcome of renal biopsies from expanded criteria donors. *Kidney Int.* **2014**, *85*, 1161–1168. [CrossRef]
35. Haas, M. Donor kidney biopsies: pathology matters, and so does the pathologist. *Kidney Int.* **2014**, *85*, 1016–1019. [CrossRef]
36. Girolami, I.; Gambaro, G.; Ghimenton, C.; Beccari, S.; Caliò, A.; Brunelli, M.; Novelli, L.; Boggi, U.; Campani, D.; Zaza, G.; et al. Pre-implantation kidney biopsy: value of the expertise in determining histological score and comparison with the whole organ on a series of discarded kidneys. *J. Nephrol.* **2020**, *33*, 167–176. [CrossRef] [PubMed]

Article

Karpinski Score under Digital Investigation: A Fully Automated Segmentation Algorithm to Identify Vascular and Stromal Injury of Donors' Kidneys

Massimo Salvi [1,*], Alessandro Mogetta [1], Kristen M. Meiburger [1], Alessandro Gambella [2], Luca Molinaro [3], Antonella Barreca [3], Mauro Papotti [4] and Filippo Molinari [1]

1 Department of Electronics and Telecommunications, Polytechnic of Turin, 10129 Turin, Italy; mogettaalessandro@gmail.com (A.M.); kristen.meiburger@polito.it (K.M.M.); filippo.molinari@polito.it (F.M.)
2 Pathology Unit, Department of Medical Sciences, University of Turin, 10126 Turin, Italy; alessandro.gambella@unito.it
3 Division of Pathology, A.O.U. Città della Salute e della Scienza Hospital, 10126 Turin, Italy; luca.molinaro@unito.it (L.M.); antonella.barreca@libero.it (A.B.)
4 Department of Oncology, University of Turin, 10126 Turin, Italy; mauro.papotti@unito.it
* Correspondence: massimo.salvi@polito.it

Received: 14 September 2020; Accepted: 3 October 2020; Published: 8 October 2020

Abstract: In kidney transplantations, the evaluation of the vascular structures and stromal areas is crucial for determining kidney acceptance, which is currently based on the pathologist's visual evaluation. In this context, an accurate assessment of the vascular and stromal injury is fundamental to assessing the nephron status. In the present paper, the authors present a fully automated algorithm, called RENFAST (Rapid EvaluatioN of Fibrosis And vesselS Thickness), for the segmentation of kidney blood vessels and fibrosis in histopathological images. The proposed method employs a novel strategy based on deep learning to accurately segment blood vessels, while interstitial fibrosis is assessed using an adaptive stain separation method. The RENFAST algorithm is developed and tested on 350 periodic acid–Schiff (PAS) images for blood vessel segmentation and on 300 Massone's trichrome (TRIC) stained images for the detection of renal fibrosis. In the TEST set, the algorithm exhibits excellent segmentation performance in both blood vessels (accuracy: 0.8936) and fibrosis (accuracy: 0.9227) and outperforms all the compared methods. To the best of our knowledge, the RENFAST algorithm is the first fully automated method capable of detecting both blood vessels and fibrosis in digital histological images. Being very fast (average computational time 2.91 s), this algorithm paves the way for automated, quantitative, and real-time kidney graft assessments.

Keywords: kidney transplantation; digital pathology; deep learning; kidney fibrosis; blood vessel segmentation; convolutional neural networks

1. Introduction

Kidney allograft transplant is experiencing a broad revolution, thanks to an increasing understanding of the pathologic mechanisms behind rejection and the introduction of new techniques and procedures for transplants [1]. The primary focus during kidney transplants has always been the identification, assessment, and treatment of allograft rejection. However, recently, a new issue has come to light: a shortage of donor organs. To solve this impasse, selection criteria were revised, leading to the so-called "expanded criteria donor" approach: kidneys that once would have been excluded because of the donors' clinical history or those deriving from deceased patients are nowadays carefully used [2,3].

In this context, the preimplantation evaluation of donors' kidneys has become more and more crucial. The pathologist's challenge is to recognize early signs of degeneration to "predict" the organs' functionality and performance. This analysis, usually based on periodic acid–Schiff (PAS) and trichrome (TRIC) staining, is focused on the glomeruli, tubules, vessels, and cortical parenchyma of the donor kidney, searching for glomerulosclerosis, tubule atrophy, vascular damage, or interstitial fibrotic replacement, respectively (Figure 1). The Karpinski score is then applied to grade the injury of the donor kidney. This score is based on a semiquantitative evaluation of the structures mentioned above. For each of the four compartments (glomeruli, tubules, blood vessels, and cortical parenchyma), the pathologist summarizes the evaluation in a four-grade score, ranging from 0 (absence of injury) to 3 (marked injuries); the total score is expressed out of 12 [4]. Notably, both arteries and arterioles are considered in vascular damage assessment, characterized by progressive thickening of their wall and shrinkage of their lumen. At the same time, the cortical parenchyma could be replaced by fibrous connective tissue [5,6].

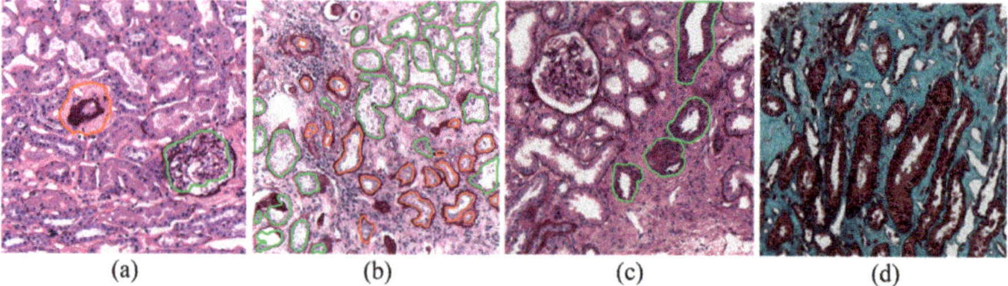

Figure 1. Histological features assessed to determine the Karpinski score. (**a**) Glomerulosclerosis: examples of a healthy and sclerotic glomerulus are shown in green and red, respectively; (**b**) Tubular atrophy: healthy and atrophic tubules are highlighted in green and red, respectively; (**c**) Vascular damage: blood vessels are outlined in green; (**d**) Cortical parenchyma: renal fibrosis is represented by the turquoise zone.

The preimplantation kidney evaluation is a delicate, crucial activity for pathology laboratories. It is time-consuming, usually performed with urgency, and has a marked impact on the daily diagnostic routine. Moreover, the evaluation is operator-dependent, with a significant rate of inter-observer difference [7]. In this challenging and evolving panorama, the introduction and application of an automated analysis algorithm would be of compelling importance.

In the last few years, several strategies have been proposed for the segmentation of kidney blood vessels and for the quantification of fibrotic tissue in biopsy images. Bevilacqua et al. [8] employed an artificial neural network (ANN) to detect blood vessels in histological kidney images. Lumen regions were firstly detected by applying fixed thresholding and morphological operators. Seeded region growing was then implemented to extract the membrane all around the segmented objects. Finally, a neural network based on Haralick texture features [9] was used to distinguish between blood vessels and tubular structures. Although well structured, this strategy suffers from several limitations. First, blood vessels with small or absent lumen cannot be segmented using the described approach. In addition, stain variability greatly influences the performance of the region growing, causing imprecise recognition of the blood vessel borders. Finally, the high variability in the shapes, dimensions, and textural characteristics of tubules seriously affects the classification provided by the network. Tey et al. [10] proposed an algorithm for the quantification of interstitial fibrosis (IF) based on color image segmentation and tissue structure identification in biopsy samples stained with Massone's trichrome (TRIC). All the renal structures were identified by employing color space transformations and structural feature extraction from the images. Then, the regions of fibrotic tissue were identified by removing all the non-fibrotic structures from the biopsy tissue area. This approach leads to fast identification of renal fibrotic tissue, but it is not free from limitations. First of all, there is a loss of information during the color space transformation and, in the presence of high stain variability, the method is not able to correctly classify all the renal structures. Moreover, being based on the identification and subsequent removal of non-fibrotic regions from the tissue, an error in the segmentation of these structures causes inaccurate quantification of interstitial fibrosis. Fu et al. [11] proposed a convolutional neural network (CNN) for fibrotic tissue segmentation in atrial tissue stained with Massone's trichrome. The network, consisting of 11 convolutional layers, was trained on a three-class problem (background vs. fibrosis vs. myocytes), giving the RGB image as input and the corresponding manual mask as the target. This approach provides fast detection of fibrotic areas of the tissue but presents one major limitation: color variability. Stain variations may affect both the training of the network and the correct segmentation of fibrotic tissue, and every mis-segmentation error leads to incorrect detection and quantification of interstitial fibrosis.

In this paper, we present a novel method for the detection of blood vessels and for the quantification of interstitial fibrosis in kidney histological images. To the best of our knowledge, no automated solution has been proposed so far to cope with the issue of stain variability in PAS and TRIC images. Our approach employs a preprocessing stage specifically designed to address the problem of color variability. The proposed algorithm for the segmentation of vascular structures exploits a deep learning approach combined with the detection of cellular structures to accurately segment blood vessels in PAS stained images. Interstitial fibrosis is assessed using an adaptive stain separation method to detect all the fibrotic areas within the histological tissue.

2. Materials and Methods

Here we present an automated method called RENFAST (Rapid EvaluatioN of Fibrosis And vesselS Thickness). The RENFAST algorithm is a deep-learning-based method for the segmentation of renal blood vessels and fibrosis. A flowchart of the proposed method is sketched in Figure 2. In the following sections, a detailed description of the algorithm is provided.

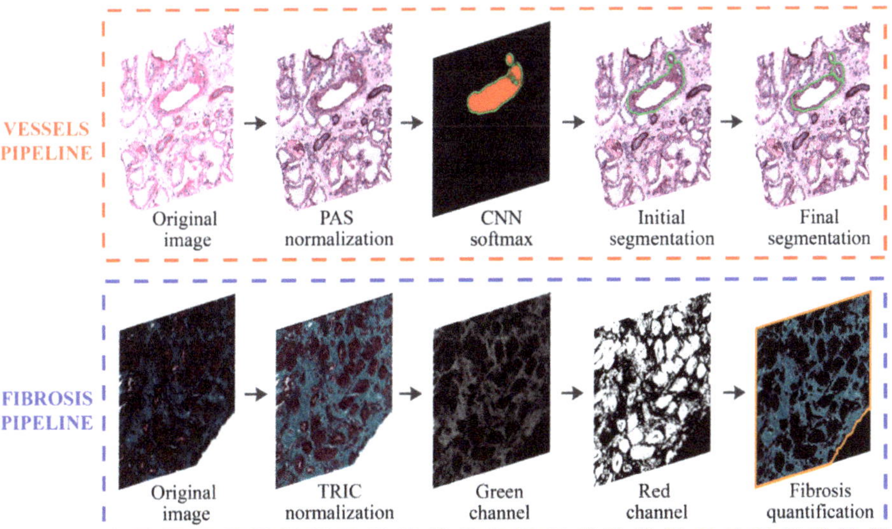

Figure 2. Flowchart of the RENFAST (Rapid EvaluatioN of Fibrosis And vesselS Thickness) algorithm for vessel and fibrosis segmentation. The first row illustrates the pipeline for blood vessel detection, while the second row shows the workflow of fibrosis segmentation. After PAS (periodic acid–Schiff) color normalization, blood vessels are detected using a deep learning method (CNN) and ad hoc post-processing. Kidney fibrosis is segmented through TRIC (Massone's trichrome) normalization followed by adaptive stain separation.

2.1. Database Description

The whole slide images (WSIs) of kidney biopsy specimens of 65 patients (median age 51 years, range 29–74 years) were used for this work; these were collected at the Division of Pathology, AOU Città della Salute e della Scienza Hospital, Turin, Italy and then anonymized. The pathology laboratory managed the biopsied samples of each kidney according to the kidney transplant biopsy's internal protocol. The tissue was fixed with Serra fixative and then processed in an urgency regimen using a microwave processor or LOGOS J processor (Milestone, Bergamo, Italy). Samples were then paraffin-embedded and serially sectioned (5 μm), mounted onto adhesive slides, and stained with PAS and TRIC. Finally, all the slides produced were scanned with a Hamamatsu NanoZoomer S210 Digital slide scanner (Turin, Italy), providing a magnification of ×100 (conversion factor: 0.934 μm/pixel). For each patient (n = 65), an expert pathologist (A.B.) manually extracted 10 images with dimensions of 512 × 512 pixels, for a total of 650 images. After consensus, manual annotations of blood vessels and fibrosis were generated by two operators (A.G. and L.M.). Table 1 shows the overall dataset composition. The image dataset, along with the annotations, is available at https://data.mendeley.com/datasets/m2t49zf6xr/1.

Table 1. Dataset used in this work.

Dataset	Subset	Stain	# Patients	# Images
Vessels	TRAIN	PAS	30	300
	TEST	PAS	5	50
Fibrosis	TRAIN	TRIC	25	250
	TEST	TRIC	5	50

2.2. Stain Normalization

The proposed algorithm employs a specific preprocessing stage, called stain normalization, to reduce the color variability of the histological samples. Previous studies have shown that stain variability significantly affects the performance of automatic algorithms in digital pathology [12,13]. The procedure of stain normalization allows for transforming a source image I into another image I_{NORM}, through the operation $I_{NORM} = f(I, I_{REF})$, where I_{REF} is a reference image and $f(\cdot)$ is the function that applies the color intensities of I_{REF} to the source image [14]. The reference image is chosen by the pathologist as the image with the most optimal tissue staining and visual appearance. For each image of the dataset, the RENFAST algorithm applies the same stain normalization method that we developed in our previous work [15]. First, the image is converted to the optical density space (OD) where the relationship between stain concentration and light intensity is linear. The algorithm then estimates the stain color appearance matrix (W) and the stain density map (H) for both the source and reference images. In order to apply the normalization, the stain density map of the source image is adjusted using the following equation:

$$I_{NORM} = W_{REF} \cdot \frac{H_{SOURCE}}{H_{REF}} \tag{1}$$

where $(\cdot)_{SOURCE}$ and $(\cdot)_{REF}$ denote the source and reference images, respectively. Finally, the normalized image is converted back from the OD space to RGB. Figure 3 illustrates the color normalization process for sample PAS and TRIC images.

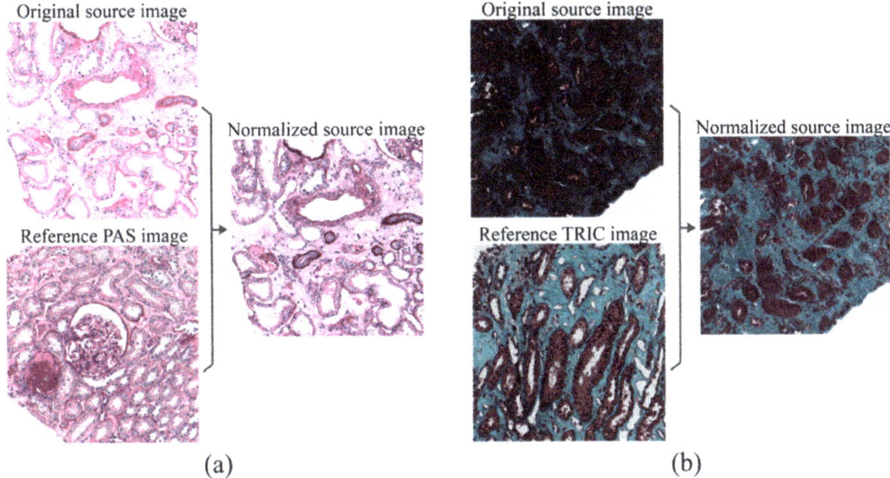

Figure 3. Stain normalization performed by the RENFAST algorithm. (**a**) PAS normalization; (**b**) TRIC normalization.

2.3. Deep Network Architecture

After stain normalization, the first step performed by the RENFAST algorithm is semantic segmentation using a convolutional neural network (CNN). To perform blood vessel segmentation, a UNET architecture with ResNet34 backbone [16] is employed using the Keras framework. The overall network architecture is shown in Figure 4. This network consists of an encoder structure that downsamples the spatial resolution of the input image through convolutional operations, to obtain a low-resolution feature mapping. These features are then resampled by a decoding structure to obtain a pixel-wise prediction of the same size of the input image. The output of the network is a probability map that assigns to each pixel a probability of belonging to a specific class. The entire network is trained on a three-class problem, giving the 512 × 512 RGB images as input and the corresponding labeled masks as the target. In each image of the dataset, pixels are labeled in three classes: (i) background, (ii) blood vessel, and (iii) blood vessel boundaries. To solve the problem of class imbalance, our network's loss function is class-weighted by taking into account how frequently a class occurs in the training set. This means that the least-represented class will have a greater contribution than a more represented one during the weight update. The class weight is computed as follows:

$$f_{classX} = \sum_{i=1}^{N} \frac{\%\, pixel_{classX}}{N} \qquad x = 1,2,3 \tag{2}$$

$$class_{WEIGHT} = \frac{median([f_{class1}, f_{class2}, f_{class3}])}{[f_{class1}, f_{class2}, f_{class3}]} \tag{3}$$

where N is the total number of images and f_{classX} is the class frequency of generic class X.

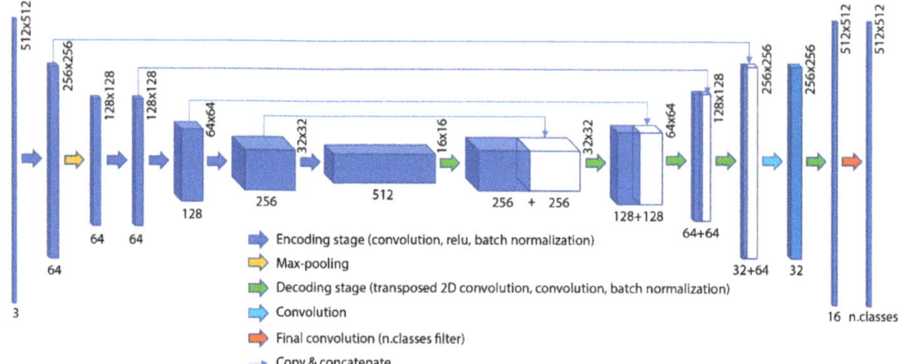

Figure 4. Architecture of the deep network employed to perform blood vessel detection. A UNET with ResNet34 backbone was implemented using Keras framework.

The encoding network was pre-trained on ILSVRC 2012 ImageNet [17]. During the training process, only the decoder weights were updated, while the encoder weights were set to non-trainable. This strategy allows for exploiting the knowledge acquired from a previous problem (ImageNet) and using the features learned to solve a new problem (vessel segmentation). This approach is useful both to speed up the training process and to create a robust model even using fewer data. The training data are real-time augmented while passing through the network, applying the same random transformations (rotation, shifting, flipping) both to the input image and to the corresponding encoded mask. Real-time data augmentation allows us to increase the amount of data available without storing the transformed data in memory. This strategy makes the model more robust to slight variations and prevents the network from overfitting.

Our network (Figure 4) was trained on 300 images with a mini-batch size of 32 and categorical cross-entropy as a loss function. The Adam optimization algorithm was employed with an initial learning rate of 0.01. The maximum number of epochs was set to 50, with a validation patience of 10 epochs for early stopping of the training process.

To preserve the information near the boundaries of the image, the RENFAST algorithm applies a specific procedure to build the CNN softmax. Briefly, a mirror border is synthesized in each direction and a sliding window approach is employed to build the probability map. To give the reader the opportunity to observe the entire procedure, we added a detailed description along with a summary figure in Appendix A.

2.4. Blood Vessel Detection

Starting from the normalized RGB image (Figure 5a), the RENFAST algorithm applies the deep network described in the previous section. Figure 5b shows the probability map obtained from the CNN, in which the red and green areas represent the pixels inside and on the edge of the blood vessels, respectively. Then, our method detects all the white and nuclear regions within the image. All the unstained structures are segmented by thresholding the grayscale image of the PAS sample, while cell nuclei are detected using the object-based thresholding developed in our previous work [15]. Figure 5c illustrates the segmentation of cellular structures performed by the RENFAST algorithm.

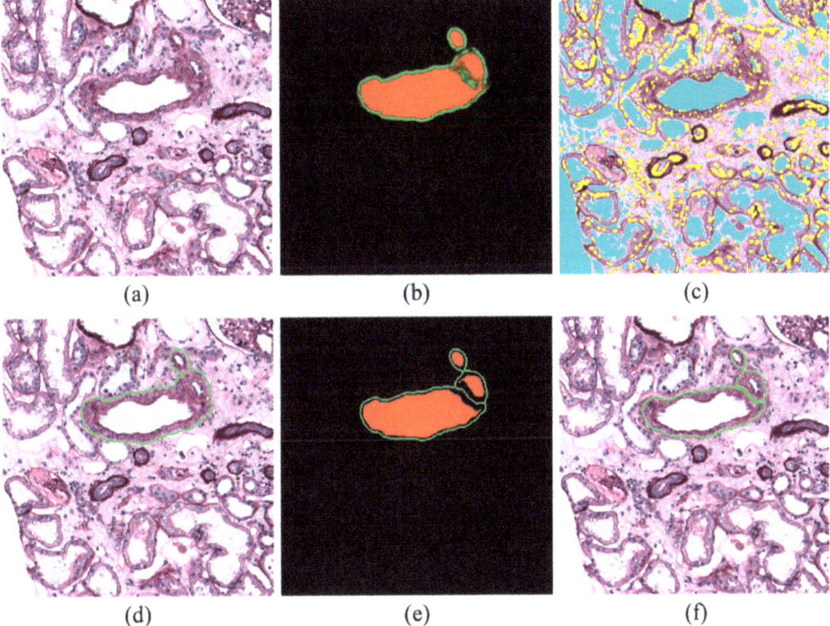

Figure 5. Steps performed by RENFAST for blood vessel detection. (**a**) Normalized image; (**b**) CNN probability map; (**c**) Cellular structure detection (yellow: nuclei, cyan: lumen); (**d**) Initial blood vessel segmentation; (**e**) Softmax with high SNR (signal-to-noise ratio); (**f**) Final blood vessel segmentation.

To obtain initial detection of the vascular structures, the probability maps of the regions inside and on the border of the blood vessels are added together and thresholded with a fixed value of 0.35. Then, morphological closing with a disk of 3-pixel radius (equal to 2.80 μm) is carried out to obtain smoother contours. As can be seen from Figure 5d, this strategy leads to accurate detection of the blood vessel boundaries but does not allow the separation of touching structures. To overcome this problem, an additional processing stage is performed to divide clustered blood vessels. The RENFAST algorithm employs a four-step procedure to increase the contrast between each blood vessel's boundary and the background:

1. Inner region mask: thresholding (0.35) and level-set on the probability map of inner regions (red layer);
2. Boundary mask: thresholding (0.35) and level-set on the probability map of boundary regions (green layer);
3. New red layer of the softmax: subtraction of the boundary mask from the inner region mask;
4. New green layer of the softmax: skeleton of the boundary mask.

This procedure generates a softmax with a high SNR (signal-to-noise ratio) where the border of each blood vessel is clearly defined (Figure 5e). Finally, for each connected component of the initial mask (Figure 5d), a simple check is performed: if by subtracting the green layer of the high-SNR softmax (Figure 5e), more than one region is generated, these regions are dilated by 1 pixel and added to the final mask. In this way, the thickness lost during the subtraction is recovered while maintaining the blood vessels' separation. Otherwise, if no additional structure is created with the subtraction, the connected component is inserted directly into the final mask.

The last step of the RENFAST algorithm for vessel segmentation is a structural check on the segmented objects: All the regions with an area less than 180 μm^2 are erased as they are too small to be considered blood vessels. In addition, objects must have at least 2.5% and 5% of the area occupied by lumen and nuclei, respectively. With these structural checks, most of the false positives generated by the CNN are deleted. The final result provided by the proposed algorithm is shown in Figure 5f.

2.5. Fibrosis Segmentation

The RENFAST algorithm is also able to quantify interstitial fibrosis in TRIC images. After stain normalization (Section 2.2), our method detects all the uncolored regions to process only TRIC stained structures. The normalized TRIC image is first converted to grayscale and Weiner filtered. The resulting image is then thresholded using a value equal to 90% of the image maximum (Figure 6a). Since fibrosis is characterized by a greenish color, the proposed algorithm applies an adaptive stain separation as described in [15]. Thanks to the stain separation (Figure 6b), it is possible to divide the regions that may manifest fibrosis (green channel) from the structural component (red channel). Segmentation of these two channels is performed using an improved version of the MANA (Multiscale Adaptive Nuclei Analysis) algorithm [18]. After min-max scaling, custom object-based thresholding is applied to the green channel (fibrosis) and red channel obtained in the previous step. For each possible threshold point $T \in [0, 1]$, the RENFAST algorithm computes the following energy function:

$$E(T) = p_0^2 \cdot var_0 \cdot log(var_0) + p_1^2 \cdot var_1 \cdot log(var_1) \quad (4)$$

where p_0 is the probability of having intensity values lower than T, p_1 is evaluated as $1 - p_0$, while var_0 and var_1 represent the variances of the probability functions of the two classes p_0 and p_1. The threshold T associated with the maximum of the energy function E represents the optimal thresholding point. The result of green and red channel segmentation is illustrated in Figure 6c. All remaining pixels not associated with one of the binary masks (white, green, red) are included in the green or red mask based on where they have the highest intensity in the stain separation channel.

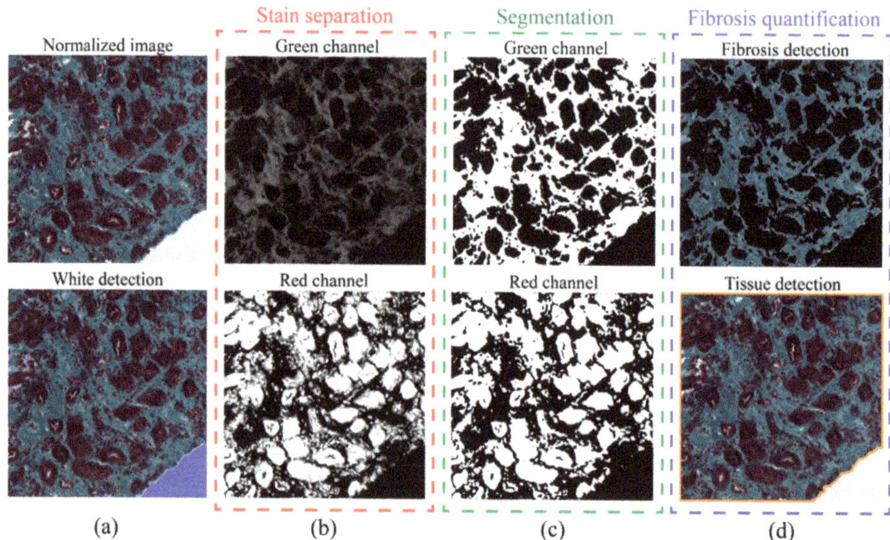

Figure 6. Steps performed by RENFAST for fibrosis segmentation. (**a**) Normalized image and white detection (in blue); (**b**) Stain separation between green and red channels; (**c**) Segmentation of green and red channels; (**d**) Fibrosis and tissue detection for interstitial fibrosis quantification.

Finally, the RENFAST algorithm quantifies the interstitial fibrosis as the ratio between the fibrotic area (segmented green channel) and the overall tissue area. Tissue detection is performed using an

RGB high-pass filter [19] where the RGB color of each pixel is treated as a 3D vector. The strength of the edge is defined as the magnitude of the maximum gradient. The raw tissue mask is generated by choosing a threshold equal to 5% of the maximum gradient. Morphological opening with a disk of 4-µm radius is then carried out to obtain the tissue contour (Figure 6d).

2.6. Performance Metrics

A comparison between manual and automatic masks was carried out to assess RENFAST's performance in the segmentation of kidney blood vessels and fibrosis. Manual annotations of blood vessels were generated using a custom graphical user interface based on MATLAB. Since fibrosis segmentation can be a long and demanding task, we designed a semi-automatic pipeline to help the pathologist during the generation of the manual mask (Appendix B). Several pixel-based metrics, such as balanced accuracy, precision, recall, and $F1_{SCORE}$, were evaluated for both blood vessel and fibrosis segmentation. Balanced accuracy ($Bal_{ACCURACY}$) is a common metric used in segmentation problems to deal with imbalanced datasets (TP vs. TN). $Bal_{ACCURACY}$ is calculated as the average of the correct predictions of each class individually. Precision is employed to evaluate the false detection of ghost shapes; recall quantifies the missed detection of ground truth objects; and finally, the $F1_{SCORE}$ is defined as the harmonic mean between precision and recall.

Accurate segmentation of blood vessel borders is fundamental for a correct evaluation of vascular damage. For this reason, we also evaluated the Dice coefficient (DSC) and the Hausdorff distance for all the true-positive vascular structures. Specifically, we computed the 95th percentile Hausdorff distance (HD95), which is defined as the maximum distance of a set (manual boundary) to the nearest point in the other set (automatic boundary). This metric is more robust towards a very small subset of outliers because it is based on the calculation of the 95th percentile of distances. During fibrosis assessment, the pathologist computes the ratio between fibrotic tissue and the whole tissue area. For each image, the absolute error (AE) between manual and automatic estimation was calculated as

$$AE = \left| \left(\frac{fibrosis_{AREA}}{tissue_{AREA}}\right)_{MANUAL} - \left(\frac{fibrosis_{AREA}}{tissue_{AREA}}\right)_{RENFAST} \right| \tag{5}$$

where $(\cdot)_{MANUAL}$ and $(\cdot)_{RENFAST}$ denote the manual and the automatic annotations, respectively.

3. Results

The automatic results provided by the RENFAST method are compared herein both with manual annotations and with previously published works. For blood vessel segmentation, we compared our algorithm with the one proposed by Bevilacqua et al. [8], while we used the methods published by Tey et al. [10] and Fu et al. [11] as benchmarks for interstitial fibrosis segmentation. As datasets and manual annotations of these works are not publicly available, all the described methods were applied to the same dataset used in this paper. The processing was performed on a custom workstation with a 3.5 GHz 10-core CPU with 64 Gb of RAM (Turin, Italy).

3.1. Blood Vessel Detection

Both pixel-based metrics ($Bal_{ACCURACY}$, precision, recall, $F1_{SCORE}$) and object-based metrics (DSC, HD95) were calculated to assess the performance of the RENFAST algorithm. To demonstrate the superiority of our strategy, we also evaluated the results obtained using a simple two-class CNN (background vs. vessel) and a three-class CNN without our post-processing. Tables 2 and 3 summarize the metrics calculated for blood vessel detection.

Table 2. Comparison between the RENFAST algorithm and the current state of the art for blood vessel segmentation (pixel-based metrics).

Method	Subset	Comp. Time (s)	$Bal_{ACCURACY}$	Precision	Recall	$F1_{SCORE}$
Bevilacqua et al. [8]	TRAIN	2.58 ± 1.24	0.6845 ± 0.1467	0.8618 ± 0.1955	0.5115 ± 0.2196	0.5996 ± 0.1931
	TEST	2.64 ± 1.18	0.6487 ± 0.1494	0.7677 ± 0.2647	0.4944 ± 0.2241	0.5684 ± 0.2281
Two-class CNN [1]	TRAIN	**0.57 ± 0.11**	0.8821 ± 0.1116	0.9203 ± 0.0945	0.8026 ± 0.1630	0.8430 ± 0.1242
	TEST	**0.56 ± 0.09**	0.8116 ± 0.1305	0.9308 ± 0.1004	0.6923 ± 0.1743	0.7741 ± 0.1419
Three-class CNN [2]	TRAIN	0.74 ± 0.16	0.8744 ± 0.0861	**0.9888 ± 0.0337**	0.7706 ± 0.1199	0.8601 ± 0.0919
	TEST	0.71 ± 0.18	0.8220 ± 0.1075	**0.9800 ± 0.0800**	0.6666 ± 0.1837	0.7740 ± 0.1597
RENFAST algorithm	TRAIN	2.67 ± 0.41	**0.9443 ± 0.0821**	0.9185 ± 0.0634	**0.9151 ± 0.0950**	**0.9126 ± 0.0611**
	TEST	2.59 ± 0.53	**0.8936 ± 0.0969**	0.9269 ± 0.0845	**0.8185 ± 0.1344**	**0.8593 ± 0.0858**

[1] CNN with the same architecture shown in Figure 2 but trained on two classes (background vs. vessel). [2] Same deep network of the RENFAST algorithm but without post-processing (Section 2.4).

Table 3. Object-based metrics calculated on detected blood vessels for both the TRAIN and TEST sets.

Method	Subset	DSC	HD95 (μm)
Bevilacqua et al. [8]	TRAIN	0.7476 ± 0.1517	20.33 ± 21.67
	TEST	0.7668 ± 0.1381	22.31 ± 34.62
Two-class CNN [1]	TRAIN	0.7447 ± 0.2312	21.13 ± 30.59
	TEST	0.6879 ± 0.2417	26.68 ± 36.50
Three-class CNN [2]	TRAIN	0.7802 ± 0.1777	12.02 ± 22.45
	TEST	0.7483 ± 0.1790	9.35 ± 8.84
RENFAST algorithm	TRAIN	**0.8441 ± 0.1762**	**9.78 ± 10.51**
	TEST	**0.8358 ± 0.1391**	**6.41 ± 6.25**

[1] CNN with the same architecture shown in Figure 2 but trained on two classes (background vs. vessel). [2] Same deep network of the RENFAST algorithm but without post-processing (Section 2.4).

Regarding pixel-based metrics, our method achieved the best $Bal_{ACCURACY}$, recall, and $F1_{SCORE}$ for both the TRAIN and TEST sets. A large margin was achieved by RENFAST compared to the state-of-the-art techniques. Even more interesting, the post-processing adopted for blood vessel segmentation allowed a further increase in the overall performance of the single deep network (three-class CNN vs. RENFAST). The combination of the CNN probability map and cellular structure segmentation increased the DSC by up to 14.8% with respect to other methods. The accurate segmentation of blood vessel boundaries is also demonstrated by the lower HD95 value. Figure 7 shows a visual comparison between RENFAST and previously published works. Our approach managed to separate and correctly outline the boundaries of the blood vessels.

3.2. Fibrosis Segmentation

The same pixel-based metrics employed in the last section were calculated to evaluate the performance of RENFAST in fibrosis quantification (Table 4). To demonstrate the importance of the stain normalization as a preprocessing step, we also evaluated the performance of our algorithm without normalizing the images ("No norm.").

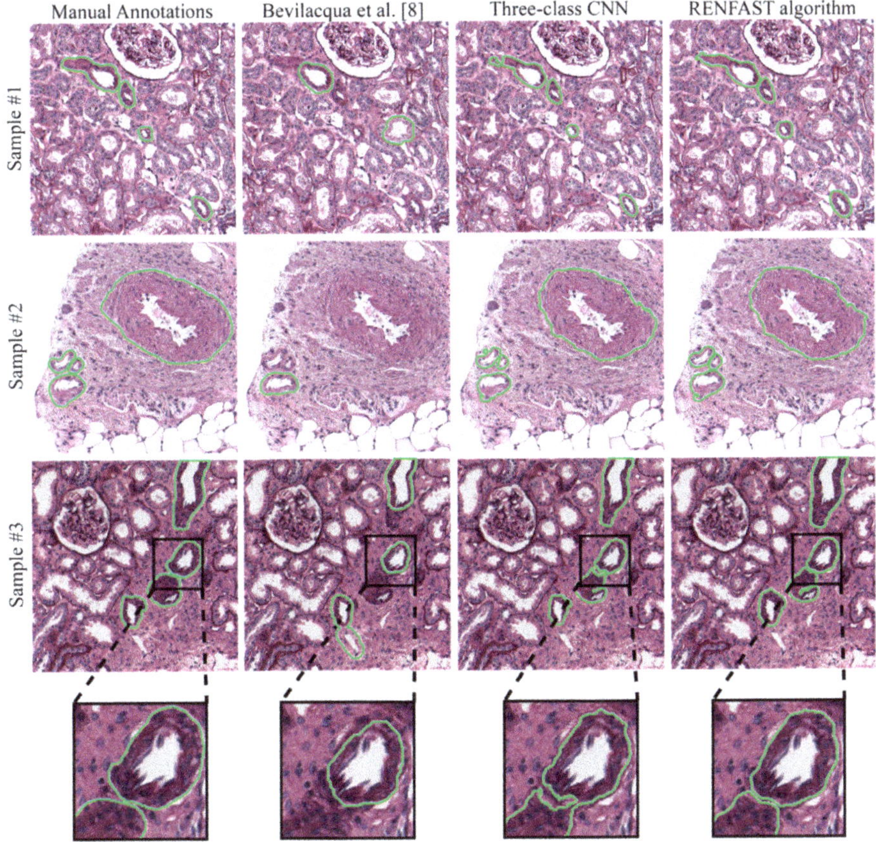

Figure 7. Blood vessel detection performed by state-of-the-art methods and the proposed algorithm. Two different samples are displayed in the first rows, while the last row shows a zoom of the segmentation near the blood vessel contour.

Table 4. Comparison between the proposed algorithm and the current state of the art for fibrosis segmentation (pixel-based metrics).

Method	Subset	Comp. Time (s)	$Bal_{ACCURACY}$	Precision	Recall	$F1_{SCORE}$
Tey et al. [10]	TRAIN	0.24 ± 0.04	0.8575 ± 0.0374	0.7538 ± 0.0780	0.8905 ± 0.0744	0.8147 ± 0.0515
	TEST	0.25 ± 0.07	0.8604 ± 0.0428	0.7512 ± 0.0736	0.9055 ± 0.0734	0.8166 ± 0.0492
Fu et al. [11]	TRAIN	**0.16 ± 0.06**	0.8988 ± 0.0660	0.8832 ± 0.1072	0.8940 ± 0.1469	0.8727 ± 0.0896
	TEST	**0.18 ± 0.09**	0.9159 ± 0.0491	0.8783 ± 0.1019	**0.9239 ± 0.1026**	0.8911 ± 0.0644
No norm. [1]	TRAIN	0.17 ± 0.07	0.9128 ± 0.0221	0.9025 ± 0.0482	0.8765 ± 0.0434	0.8900 ± 0.0240
	TEST	0.18 ± 0.11	0.9164 ± 0.0247	0.9157 ± 0.0304	0.8738 ± 0.0499	0.8944 ± 0.0277
RENFAST algorithm	TRAIN	0.27 ± 0.13	**0.9212 ± 0.0199**	**0.9064 ± 0.0355**	**0.8958 ± 0.0480**	**0.8973 ± 0.0275**
	TEST	0.29 ± 0.14	**0.9227 ± 0.0222**	**0.9184 ± 0.0313**	0.8891 ± 0.0482	**0.9010 ± 0.0246**

[1] RENFAST algorithm without the stain normalization as preprocessing.

As shown in Table 4, our strategy outperformed all the previously published methods. In addition, the stain normalization (Section 2.2) allowed a further increase in the overall performance of our method (No norm. vs. RENFAST algorithm). Finally, we evaluated the absolute errors (AEs) between the manual and automatic fibrosis quantification (Table 5). In both the TRAIN and TEST datasets, the RENFAST algorithm achieved the lowest average AEs (2.42% and 2.32%), with maximum AEs of 11.17% and 7.81%, respectively. Specifically, the maximum AE obtained by our method was 3–5 times lower compared to state-of-the-art techniques [10,11]. Figure 8 shows some kidney fibrosis segmentation results.

Table 5. Minimum, average, and maximum absolute errors (AE_{MIN}, AE_{MEAN}, AE_{MAX}) between manual and automatic fibrosis quantification.

Method	Subset	AE_{MIN} (%)	AE_{MEAN} (%)	AE_{MAX} (%)
Tey et al. [10]	TRAIN	0.03	8.79	42.46
	TEST	0.59	8.73	38.41
Fu et al. [11]	TRAIN	0.01	7.81	38.62
	TEST	0.04	5.93	28.73
No norm. [1]	TRAIN	0.01	2.52	11.21
	TEST	0.05	2.50	8.29
RENFAST algorithm	TRAIN	**0.01**	**2.42**	**11.17**
	TEST	**0.01**	**2.32**	**7.81**

[1] RENFAST algorithm without the stain normalization as preprocessing.

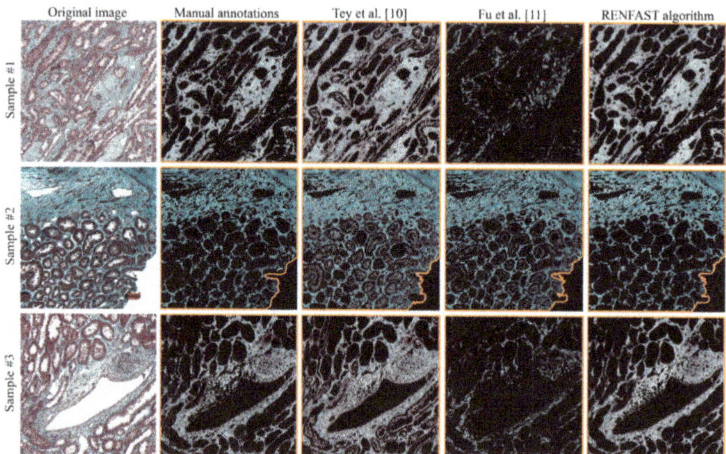

Figure 8. Visual performance comparison between previously published papers for fibrosis detection and the RENFAST algorithm. The fibrosis mask is superimposed on the original image, while the tissue contour is highlighted in orange.

3.3. *Whole Slide Analysis*

Since arteriosclerosis and fibrosis are generally assessed on whole slide images (WSIs), we extended our strategy to entire biopsies using a sliding window approach. To evaluate the degree of arterial sclerosis and fibrosis, an expert pathologist takes at least 20 min per patient, while the RENFAST algorithm is able to process the entire WSI in about 2 min. Figure 9 illustrates the results obtained using our algorithm on two different kidney biopsies stained with PAS (vessel detection) and TRIC (fibrosis segmentation). The introduction of an automatic algorithm within the clinical workflow can speed up the diagnostic process and provide more accurate data to assess kidney transplantability.

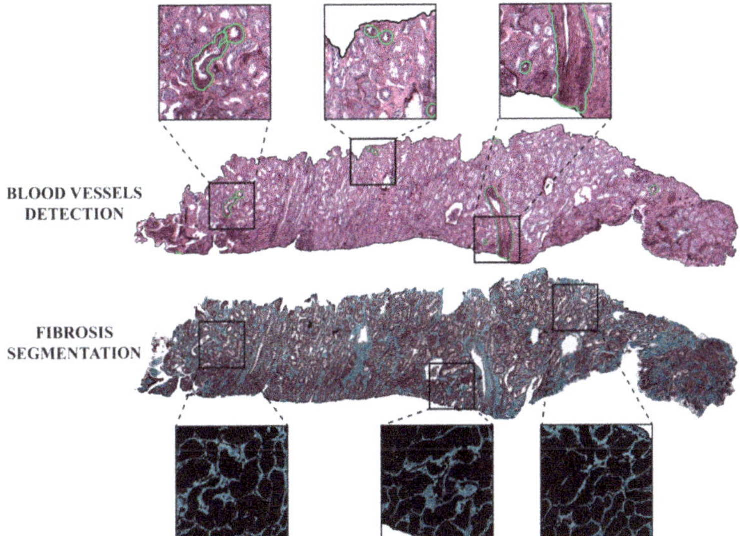

Figure 9. The result of RENFAST processing on a whole slide image (WSI). Blood vessels are shown in green in PAS stained WSIs. During the assessment of fibrosis, the connective tissue is segmented by removing all the tubular, vascular, and glomerular structures.

4. Discussion and Conclusions

Advances in transplant patient management are steadily increasing with improved clinical data and outcomes, requiring proportional development of the technical procedures routinely applied. However, the histopathological evaluation of preimplantation donor kidney biopsies has not varied, despite the increasing demand for pathology reports.

In this study, we present a fast and accurate method for the segmentation of kidney blood vessels and fibrosis in histological images. The detection of vascular structures and interstitial fibrosis is a real challenge due to the stain variability that affects the PAS and TRIC images, combined with high variation in the shape, size, and internal architecture of the renal structures. Thanks to the stain normalization step, our approach is capable of automatically detecting fibrotic areas and blood vessels in images with different staining intensity. The proposed algorithm was developed and tested on 350 PAS images for blood vessel segmentation and on 300 TRIC stained images for the detection of renal fibrosis. The results were compared with both manual annotations and previously published methods [8,10,11].

In blood vessel detection, the RENFAST algorithm achieved the best $Bal_{ACCURACY}$, recall, and $F1_{SCORE}$ compared to other techniques. More importantly, our strategy obtained the best DSC and HD95 in the segmentation of vessel boundaries (Table 3). This is fundamental as accurate segmentation of the blood vessel borders is mandatory for the correct evaluation of vascular damage. This high performance is mainly due to the combination of CNN segmentation with ad hoc post-processing specifically designed to detect the contour of each blood vessel. By segmenting lumen regions and cell nuclei, the RENFAST algorithm manages to delete almost all the false-positive shapes detected by the CNN. Our strategy is also capable of segmenting small blood vessels and correctly separating touching structures (Figure 7).

On TRIC stained images, the RENFAST algorithm allows us to quantify the interstitial fibrosis. The proposed approach showed high accuracy in segmenting fibrotic tissue and outperformed all the previously published methods (Table 4). Compared with the current state-of-the-art techniques, our method obtained the lowest absolute error (around 2.4%) in the estimation of fibrosis percentage.

In the TEST set, the maximum absolute error of the algorithm was only 7.81%, about 4 times lower with respect to the compared methods. The combination of color normalization and adaptive stain separation allows us to accurately quantify the extent of the fibrotic area.

Although the proposed strategy is fast and robust, it still has some limitations. First of all, the histological images must be acquired at 10× or higher magnification. Using a lower resolution (5× or below), the deep network cannot accurately segment the blood vessels, and cell nucleus segmentation may fail due to the poor quality of the image. Another limitation refers to the WSI application. Nowadays, pathologists evaluate only arteriolar narrowing and interstitial fibrosis in the renal cortex, excluding all structures of the medulla from the evaluation. Our algorithm does not yet include a pipeline for the recognition of the medullary tissue from the cortical tissue on kidney biopsies. However, its potential in assessing vessel and parenchyma injury represents an efficient tool to increase accuracy, reproducibility, and velocity in an increasingly urgent medical setting.

In this study, we presented a simple yet effective pipeline for blood vessel and fibrosis segmentation in kidney histological images. Our research group is currently working on the extension of the RENFAST algorithm to automatically detect the cortical tissue on WSIs and assign a vascular score according to [5]. In the future, we will integrate the assessment of glomerulosclerosis and tubular atrophy within the RENFAST algorithm in order to create the first automated Karpinski scoring system.

Author Contributions: Conceptualization, L.M. and F.M.; methodology, M.S.; software, M.S. and A.M.; validation, A.M. and K.M.M.; resources, A.G. and A.B.; data curation, A.G. and L.M.; writing—original draft preparation, M.S.; writing—review and editing, A.M. and K.M.M.; supervision, M.P. and F.M. All authors have read and agreed to the published version of the manuscript.

Funding: This research received no external funding.

Acknowledgments: The authors would like to acknowledge all the laboratory technicians of the Division of Pathology (Department of Oncology, Turin, Italy) for their help in digitizing histological slides.

Conflicts of Interest: The authors declare no conflict of interest.

Appendix A

During the inference phase, the CNN's probability map could suffer from a lack of information near the edges of the image. To overcome this problem, an *Extended image* is synthesized by padding the original image with mirror reflections of 256 × 256 pixels along each direction. As shown in Figure A1, the result of this operation is an RGB image of 1024 × 1024 pixels. A sliding window operator with a size of 512 × 512 is then passed over the extended image with an overlap of 256 pixels between consecutive windows. The deep network is applied to each 512 × 512 window, and only the center of each prediction is kept for the creation of the initial softmax. This operation yields a heat map of size 768 × 768 which is further center cropped to obtain the final softmax with the same size as the input image. The final softmax can be considered as an RGB image, where the red layer contains the probability for each pixel of belonging to the "blood vessel" class, while the green layer represents the probability for each pixel of belonging to the "blood vessel boundaries" class.

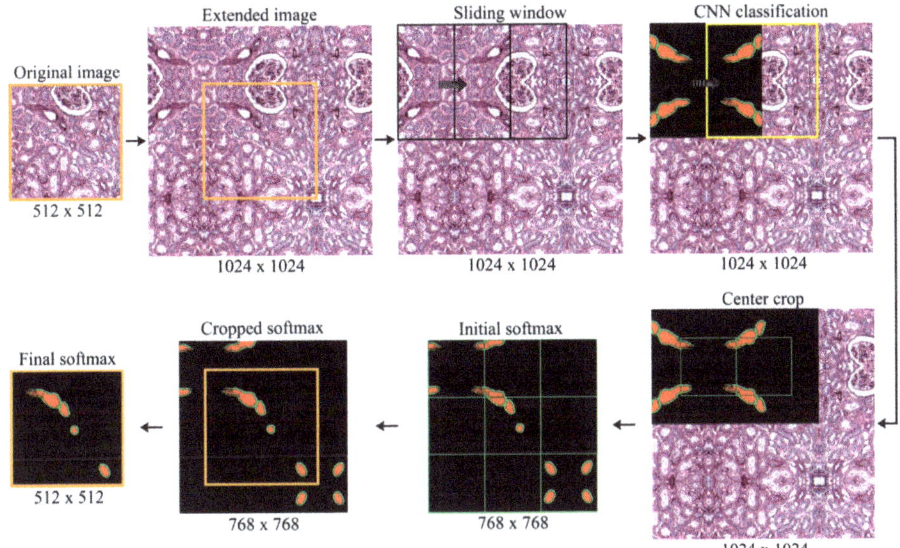

Figure A1. Procedure for the creation of the final CNN softmax. The original image is mirrored around the boundaries to obtain the extended image. Then, a sliding window approach is employed to classify each patch, and only the center of each prediction is kept to build the final softmax.

Appendix B

The semi-automatic pipeline used to generate the manual annotation of fibrotic areas was developed in Fiji [20]. Fiji is a Java-based software product with several plugins that facilitate medical image analysis. The proposed pipeline consists of seven steps: (i) image loading; (ii) manual definition of a ROI (region of interest) for each of the three colors (white, green, red); (iii) RGB color averaging of each ROI to obtain the three stain vectors; (iv) color deconvolution using the stain vectors previously found; (v) manual thresholding on the green channel; (vi) small particle removal; and (vii) complementation of the binary mask.

References

1. Salmon, G.; Salmon, E. Recent Innovations in Kidney Transplants. *Nurs. Clin. N. Am.* **2018**, *53*, 521–529. [CrossRef] [PubMed]
2. Metzger, R.A.; Delmonico, F.L.; Feng, S.; Port, F.K.; Wynn, J.J.; Merion, R.M. Expanded criteria donors for kidney transplantation. *Am. J. Transplant.* **2003**, *3*, 114–125. [CrossRef] [PubMed]
3. Heilman, R.L.; Mathur, A.; Smith, M.L.; Kaplan, B.; Reddy, K.S. Increasing the use of kidneys from unconventional and high-risk deceased donors. *Am. J. Transplant.* **2016**, *16*, 3086–3092. [CrossRef]
4. Altini, N.; Cascarano, G.D.; Brunetti, A.; Marino, F.; Rocchetti, M.T.; Matino, S.; Venere, U.; Rossini, M.; Pesce, F.; Gesualdo, L. Semantic Segmentation Framework for Glomeruli Detection and Classification in Kidney Histological Sections. *Electronics* **2020**, *9*, 503. [CrossRef]
5. Karpinski, J.; Lajoie, G.; Cattran, D.; Fenton, S.; Zaltzman, J.; Cardella, C.; Cole, E. Outcome of kidney transplantation from high-risk donors is determined by both structure and function. *Transplantation* **1999**, *67*, 1162–1167. [CrossRef] [PubMed]
6. Carta, P.; Zanazzi, M.; Caroti, L.; Buti, E.; Mjeshtri, A.; Di Maria, L.; Raspollini, M.R.; Minetti, E.E. Impact of the pre-transplant histological score on 3-year graft outcomes of kidneys from marginal donors: A single-centre study. *Nephrol. Dial. Transplant.* **2013**, *28*, 2637–2644. [CrossRef]
7. Furness, P.N.; Taub, N.; Project, C. of E.R.T.P.A.P. (CERTPAP) International variation in the interpretation of renal transplant biopsies: Report of the CERTPAP Project. *Kidney Int.* **2001**, *60*, 1998–2012. [CrossRef]

8. Bevilacqua, V.; Pietroleonardo, N.; Triggiani, V.; Brunetti, A.; Di Palma, A.M.; Rossini, M.; Gesualdo, L. An innovative neural network framework to classify blood vessels and tubules based on Haralick features evaluated in histological images of kidney biopsy. *Neurocomputing* **2017**, *228*, 143–153. [CrossRef]
9. He, D.-C.; Wang, L. Texture features based on texture spectrum. *Pattern Recognit.* **1991**, *24*, 391–399. [CrossRef]
10. Tey, W.K.; Kuang, Y.C.; Ooi, M.P.-L.; Khoo, J.J. Automated quantification of renal interstitial fibrosis for computer-aided diagnosis: A comprehensive tissue structure segmentation method. *Comput. Methods Programs Biomed.* **2018**, *155*, 109–120. [CrossRef] [PubMed]
11. Fu, X.; Liu, T.; Xiong, Z.; Smaill, B.H.; Stiles, M.K.; Zhao, J. Segmentation of histological images and fibrosis identification with a convolutional neural network. *Comput. Biol. Med.* **2018**, *98*, 147–158. [CrossRef]
12. Monaco, J.; Hipp, J.; Lucas, D.; Smith, S.; Balis, U.; Madabhushi, A. Image segmentation with implicit color standardization using spatially constrained expectation maximization: Detection of nuclei. In Proceedings of the International Conference on Medical Image Computing and Computer-Assisted Intervention, Nice, France, 1–5 October 2012; Springer: Berlin/Heidelberg, Germany, 2012; pp. 365–372.
13. Peter, L.; Mateus, D.; Chatelain, P.; Schworm, N.; Stangl, S.; Multhoff, G.; Navab, N. Leveraging random forests for interactive exploration of large histological images. In Proceedings of the International Conference on Medical Image Computing and Computer-Assisted Intervention, Boston, MA, USA, 14–18 September 2014; Springer: Cham, Switzerland, 2014; pp. 1–8.
14. Ciompi, F.; Geessink, O.; Bejnordi, B.E.; De Souza, G.S.; Baidoshvili, A.; Litjens, G.; Van Ginneken, B.; Nagtegaal, I.; Van Der Laak, J. The importance of stain normalization in colorectal tissue classification with convolutional networks. In Proceedings of the 2017 IEEE 14th International Symposium on Biomedical Imaging (ISBI 2017), Melbourne, Australia, 18–21 April 2017; pp. 160–163.
15. Salvi, M.; Michielli, N.; Molinari, F. Stain Color Adaptive Normalization (SCAN) algorithm: Separation and standardization of histological stains in digital pathology. *Comput. Methods Programs Biomed.* **2020**, *193*, 105506. [CrossRef]
16. Ronneberger, O.; Fischer, P.; Brox, T. U-net: Convolutional networks for biomedical image segmentation. In Proceedings of the International Conference on Medical Image Computing and Computer-Assisted Intervention, Munich, Germany, 5–9 October 2015; Springer: Cham, Switzerland, 2015; pp. 234–241.
17. Krizhevsky, A.; Sutskever, I.; Hinton, G.E. Imagenet classification with deep convolutional neural networks. In Proceedings of the Advances in Neural Information Processing Systems, Lake Tahoe, NV, USA, 3–8 December 2012; pp. 1097–1105.
18. Salvi, M.; Molinari, F. Multi-tissue and multi-scale approach for nuclei segmentation in H&E stained images. *Biomed. Eng. Online* **2018**, *17*. [CrossRef]
19. Salvi, M.; Molinaro, L.; Metovic, J.; Patrono, D.; Romagnoli, R.; Papotti, M.; Molinari, F. Fully automated quantitative assessment of hepatic steatosis in liver transplants. *Comput. Biol. Med.* **2020**, *123*, 103836. [CrossRef]
20. Schindelin, J.; Arganda-Carreras, I.; Frise, E.; Kaynig, V.; Longair, M.; Pietzsch, T.; Preibisch, S.; Rueden, C.; Saalfeld, S.; Schmid, B. Fiji: An open-source platform for biological-image analysis. *Nat. Methods* **2012**, *9*, 676–682. [CrossRef]

Article

Deep Learning with Limited Data: Organ Segmentation Performance by U-Net

Michelle Bardis [1,*], Roozbeh Houshyar [1], Chanon Chantaduly [1], Alexander Ushinsky [2], Justin Glavis-Bloom [1], Madeleine Shaver [1], Daniel Chow [1], Edward Uchio [3] and Peter Chang [1]

1 Department of Radiological Sciences, University of California, Irvine, CA 92617, USA; rhoushya@hs.uci.edu (R.H.); cchantad@hs.uci.edu (C.C.); jglavisb@hs.uci.edu (J.G.-B.); mshaver@hs.uci.edu (M.S.); chowd3@hs.uci.edu (D.C.); changp6@hs.uci.edu (P.C.)

2 Mallinckrodt Institute of Radiology, Washington University Saint Louis, St. Louis, MO 63110, USA; aushinsky@wustl.edu

3 Department of Urology, University of California, Orange, CA 92868, USA; euchio@hs.uci.edu

* Correspondence: mbardis@hs.uci.edu

Received: 19 June 2020; Accepted: 23 July 2020; Published: 26 July 2020

Abstract: (1) Background: The effectiveness of deep learning artificial intelligence depends on data availability, often requiring large volumes of data to effectively train an algorithm. However, few studies have explored the minimum number of images needed for optimal algorithmic performance. (2) Methods: This institutional review board (IRB)-approved retrospective review included patients who received prostate magnetic resonance imaging (MRI) between September 2014 and August 2018 and a magnetic resonance imaging (MRI) fusion transrectal biopsy. T2-weighted images were manually segmented by a board-certified abdominal radiologist. Segmented images were trained on a deep learning network with the following case numbers: 8, 16, 24, 32, 40, 80, 120, 160, 200, 240, 280, and 320. (3) Results: Our deep learning network's performance was assessed with a Dice score, which measures overlap between the radiologist's segmentations and deep learning-generated segmentations and ranges from 0 (no overlap) to 1 (perfect overlap). Our algorithm's Dice score started at 0.424 with 8 cases and improved to 0.858 with 160 cases. After 160 cases, the Dice increased to 0.867 with 320 cases. (4) Conclusions: Our deep learning network for prostate segmentation produced the highest overall Dice score with 320 training cases. Performance improved notably from training sizes of 8 to 120, then plateaued with minimal improvement at training case size above 160. Other studies utilizing comparable network architectures may have similar plateaus, suggesting suitable results may be obtainable with small datasets.

Keywords: training size; deep learning; convolutional neural network; U-Net; segmentation; artificial intelligence

---

1. Introduction

Deep learning through convolutional neural networks (CNNs), a subset of artificial intelligence, has demonstrated many strengths for image analysis [1]. For example, CNN approaches represent all recent winning entries within the annual ImageNet Classification challenge, consisting of over one million photographs in 1000 object categories with a 3.6% classification error rate to date [2,3]. In addition, medical applications have demonstrated potential to improve triage with intracranial hemorrhage detection [4] and glioma genetic mutation classification [5]. However, a CNN's performance depends on its ability to learn from the input data itself, and a CNN requires both (1) high-quality and (2) large datasets to solve problems effectively [6,7]. By determining the relationship between dataset size and CNN accuracy, investigators could potentially calculate when a CNN has been effectively trained.

Training data scarcity and quality are generally not considered challenges for non-biomedical applications where data is widely available. For example, Facebook collects more than 50 TB of video per day and Google processes 200,000 TB per day [8,9]. By contrast, biomedical datasets tend to be heterogenous, difficult to annotate, and relatively scarce [10,11]. In two recent breast imaging studies that used artificial intelligence (AI), the dataset sizes for breast lesion detection and breast cancer recurrence were 320 and 92 patients, respectively [12,13]. Medical studies often lack a combination of publicly available data and high-quality labels [1,14]. Recognition of rare diseases proves especially challenging for medical imaging neural networks, as imaging data for these diseases are often very limited [14]. Additionally, annotation of clinical data is a time consuming and potentially expensive process. Consequently, most medical imaging CNNs face a scarcity of data and calculating an optimal dataset size is infeasible [14].

Since most medical imaging studies are constrained by small datasets, few studies have examined the relationship between the number of cases and CNN performance. A study by Cho et al. [15] compared the number of cases versus performance for a CNN that classified axial computerized tomography scans (CTs) into different anatomic regions: brain, neck, shoulder, chest, abdomen, and pelvis. Another study by Lakhani et al. [16] also observed the performance difference with four different case sizes for CNNs that identified the presence or absence of an endotracheal tube on chest radiographs. Although these two studies showed better accuracy with more cases, the CNNs utilized in the studies completed image classification tasks that make a binary decision after examining the image in its entirety. The relationship between number of cases and segmentation performance within an image has not been rigorously explored.

The purpose of this study is to identify the ideal training size for prostate organ segmentation by analyzing the relationship between the number of MRI cases utilized and consequent CNN performance for imaging analysis. We implemented a type of CNN called a U-Net [17], which was specifically created for medical imaging assessment tasks typically lacking large datasets. U-Net is widely used in medical imaging artificial intelligence (AI) research. We hypothesize a plateau in performance because organ segmentation is a suitable and straightforward task for a U-Net.

2. Materials and Methods

2.1. Patient Selection

This retrospective study was granted a waiver of informed consent by the institutional review board (IRB) at the University of California, Irvine (UCI) for use of human subject data in a research study. An institutional prostate cancer database was searched to identify patients who had both a (1) prostate multiparametric magnetic resonance imaging (mpMRI) and a (2) magnetic resonance imaging/transrectal ultrasound (MRI/TRUS) fusion biopsy between September 2014 and August 2018. The inclusion criteria for this study included patients who had an mpMRI with subsequent 12-core Artemis 3D TRUS (Eigen, Grass Valley, CA, USA) and MRI/TRUS fusion biopsy using Artemis and ProFuse software (Eigen, Grass Valley, CA, USA) at the University of California, Irvine. An MRI/TRUS fusion biopsy was included as criteria because the prostate organ ground truth was segmented for these patients.

2.2. Image Acquisition

The mpMRI images were acquired on a Siemens Magnetom Trio 3-Tesla MRI scanner (Siemens AG, Munich, Germany) and a Phillips Ingenia 3-Tesla MRI scanner (Phillips Healthcare, Amsterdam, Netherlands) at the University of California, Irvine. The image acquisitions were completed in adherence to the prostate imaging reporting and data system (PI-RADS) v2 protocol without endorectal coil (Table 1).

Table 1. Magnetic resonance imaging (MRI) acquisition parameters.

Parameter	Measure
Field Strength (B0)	3 Tesla
Acquisition Technique	Turbo spin echo/echo planar
Echo Train Length	25
Time Repetition	7300 milliseconds
Time Echo	108 milliseconds
Flip Angle	150 degrees
Field of View	200 × 200 voxels
Matrix Size	256 × 205 pixels
Slice Thickness	3 mm
Slice Spacing	3 mm
Coil	Body

2.3. Ground Truth Segmentation

Ten radiologists manually segmented the prostate organ on axial T2-weighted (T2W) images with Profuse software (Eigen, Grass Valley, CA, USA). A board-certified abdominal radiologist with over 10 years of experience (R.H.) was the most experienced radiologist who approved each case. When other radiologists' segmentations differed from his expertise, he refined and updated those segmentations to establish the final ground truth. The mpMRI data and prostate organ segmentation data were transferred to a proprietary research database. From the database, the T2W axial images and organ segmentations were accessed and revised on an in-house image segmenting tool. The in-house tool enabled any segmentation corrections to be completed quickly. This tool integrated with the neural network training software and could be accessed with a web browser. Any segmentation updates were thus seamlessly updated into the neural network implementation.

2.4. Image Preprocessing

All axial images were resized to 256 × 256 voxels for neural network training. The axial slices were set to have a distance of 3 mm between each other. The standard deviation and mean values of each image were calculated when retrieved from the database. The image signal intensity was then normalized and applied voxelwise to each image. From all the available mpMRI sequences, only the T2W images were used for training and validation.

2.5. Convolutional Neural Network

The CNN used for this study was a custom modified U-Net. The algorithm's base architecture was derived from a standard U-Net, which is a fully convolutional contracting and expanding architecture [17]. The customized U-Net has a symmetric architecture and uses the same number of layers during subsampling and upsampling. U-Net also employs skip connections that allow the CNN to combine features for the image contraction and expansion pathways. These skip connections enabled the U-Net to use spatial information that could potentially be lost after the image is further downsampled in the contraction pathway. The entire image was trained during a single forward pass and the U-Net classified each image per pixel.

Our customized U-Net was extended to incorporate three dimensions during training and then produce outputs in two dimensions (Figure 1). Five layers were chosen empirically. In each layer, the image was processed by batch normalization, convolution, rectified linear unit (ReLU) activation, and downsampling with strided convolutions by a factor of 2. The 5 layers used 4, 8, 16, 32, and 32 filters per convolution. The image was downsampled until it became a 1 × 1 × 1 matrix before it underwent expansion. During the expansion pathway, the image was upsampled and a skip connection allowed the upsampled image to combine spatial information from the contraction pathway.

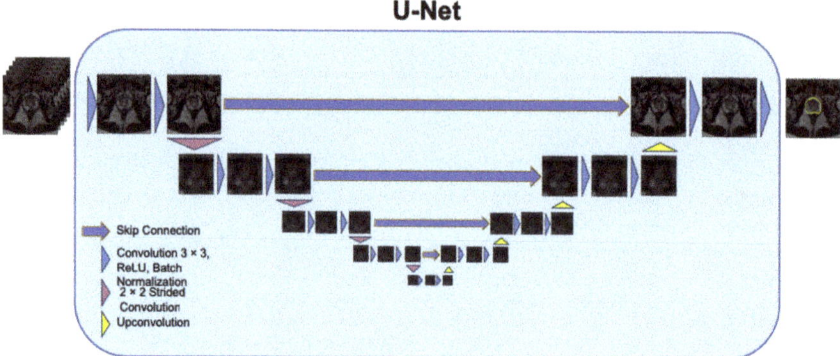

Figure 1. All neural network runs were completed on a U-Net with 5 layers. The number of channels used were 4, 8, 16, 32, and 32 for the 5 layers.

2.6. Algorithm Training

The Adam optimization algorithm was employed to update the network weights. The Adam algorithm used classical stochastic gradient descent during training [18]. The learning rate was set to 1×10^{-3}, while the exponential decay rates, β_1 and β_2, were set to 0.9 and 0.999, respectively. The batch size was set to 32. The U-Net was trained over a range of iterations: 12,000 to 96,000. The hyperparameters and network structure were kept constant across all 12 runs.

The CNN was written with TensorFlow r1.9 library (Apache 2.0 license) and Python 3.5. The neural network was trained on a graphics processing unit (GPU) workstation which employed four GeForce GTX 1080 Ti cards (11 GB, Pascal microarchitecture; NVIDIA, Santa Clara, CA, USA).

2.7. Statistical Methods

The U-Net performance was measured by examining the Dice score. X and Y are both spatial target regions and their overlap is defined by the Dice score:

$$\text{Dice} = \frac{2\,|\,X \cap Y\,|}{|X| + |Y|}. \tag{1}$$

The Dice score quantifies the spatial overlap between the manually segmented and neural network-derived segmentations (Appendix A, Figure A1). A Dice score ranges from 0 (no overlap) to 1 (perfect overlap). A Dice score is the most widely used metric for evaluating segmentation performance for a neural network [19]. To estimate the stability of the neural network during training, the variance of the training Dice score was calculated.

The total number of cases available for training and validation was 400 MRIs. Our U-Net was implemented for 12 runs and trained on the following number of cases: 8, 16, 24, 32, 40, 80, 120, 160, 200, 240, 280, and 320 cases. For each of the 12 runs, the cases were randomly partitioned as either training or validation and the entire set of 400 cases were used. The Dice score was calculated for every validation case in every run. From validation cases in every run, the mean and standard deviation of the Dice scores were computed. For example, the CNN in Run 1 was trained on 8 cases. After the CNN was done training, validation on 392 cases that produced 392 different Dice scores was completed. The mean and standard deviation for these 392 Dice scores were 0.424 and 0.206, respectively. Training size, validation size, mean Dice score, and standard deviation of Dice score are listed for Runs 1 through 12 in Table 2.

Table 2. Mean Dice score and standard deviation of Dice score for 12 training sizes.

Run	Training Size (Cases)	Validation Size (Cases)	Mean Dice Score	Standard Deviation of Dice Score
1	8	392	0.424	0.206
2	16	384	0.653	0.160
3	24	376	0.716	0.145
4	32	368	0.724	0.150
5	40	360	0.747	0.147
6	80	320	0.819	0.099
7	120	280	0.793	0.113
8	160	240	0.858	0.068
9	200	200	0.840	0.111
10	240	160	0.855	0.076
11	280	120	0.857	0.082
12	320	80	0.867	0.090

After calculating the mean Dice score for 12 different runs, the SciPy [20] library in Python was used to complete curve fitting to these 12 data points with three nonlinear functions (Equations (2)–(4)). Multiple functions were used to optimize the regression and the best three functions that approximate the data were shown (Equations (2)–(4)). These three functions were selected from the SciPy library because they most effectively modeled the dataset that increased quickly from training sizes 8 to 32 and then gradually from training sizes 200 to 320. For all three functions, a, b, and c were constants, y was the Dice score, and x was the training size (Figure 2). The first function was logarithmic, with the formula:

$$y = a \times \ln(x) + b. \quad (2)$$

The second function was asymptotic and used the formula:

$$y = \frac{a}{b + \frac{1}{x}}. \quad (3)$$

The third function was exponential, with the formula:

$$y = 1 - a \times e^{-b \times x} + c \quad (4)$$

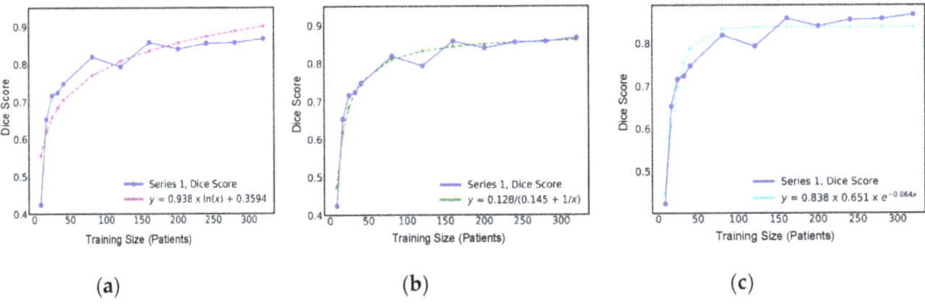

Figure 2. The mean Dice score at 12 different training sizes was approximated with several curve functions. (**a**) The first function was logarithmic with the formula $y = a \times \ln(x) + b$. a was 0.938 and b was 0.3594. The mean squared error was 2.55×10^{-3}. (**b**) The second function was asymptotic and used the formula $y = \frac{a}{b+\frac{1}{x}}$. a was 0.128 and b was 0.145. The mean squared error was 5.70×10^{-4}. (**c**) The third function was exponential with the formula $y = 1 - a \times e^{-b \times x} + c$. a was 0.651, b was 0.064, and c was −0.162. The mean squared error was 8.30×10^{-4}. The second function (**b**) provided the best approximation because it had the lowest mean squared error.

For each approximation, the mean squared error was calculated with the following formula:

$$\text{Mean Squared Error} = \frac{1}{n}\sum\nolimits_{i=1}^{n}(y_i - \widetilde{y_i})^2 \tag{5}$$

where n was 12, y was the Dice score, and $\widetilde{y_i}$ was the estimated Dice score produced by the function.

3. Results

3.1. Prostate Segmentation

A total of 400 cases (10,400 axial images) from 374 patients were used during training and validation in our study. The average patient age was 65 years (range 41 to 96 years). The average prostate volume was 59 cm^3 (range 2 cm^3 to 353 cm^3). The relationship between number of cases used for training and algorithm performance is shown in Figure 3. The Dice score improved most when the case number changed from 8 to 16 (Table 2). In addition, the Dice score also started to plateau at a training size of 160 cases. The Dice score was 0.858 at 160 cases and 0.867 at 320 cases. To show progression of the Dice score, a single axial image from one case was selected to show the benefits of increasing the number of cases (Figure 4). On this one axial slice, the Dice score progressed from 0 to 0.98 as the training size grew from 8 to 320 cases.

Three nonlinear functions from the SciPy library were used to best fit the mean Dice score performance across the 12 runs. For the first function (2), a was 0.938, b was 0.3594, and the mean squared error was 2.55×10^{-3}. For the second function (3), a was 0.128, b was 0.145, and the mean squared error was 5.70×10^{-4}. For the third function (4), a was 0.651, b was 0.064, c was −0.162, and the mean squared error was 8.30×10^{-4}. The best curve fitting was completed by the second function and produced the lowest mean squared error.

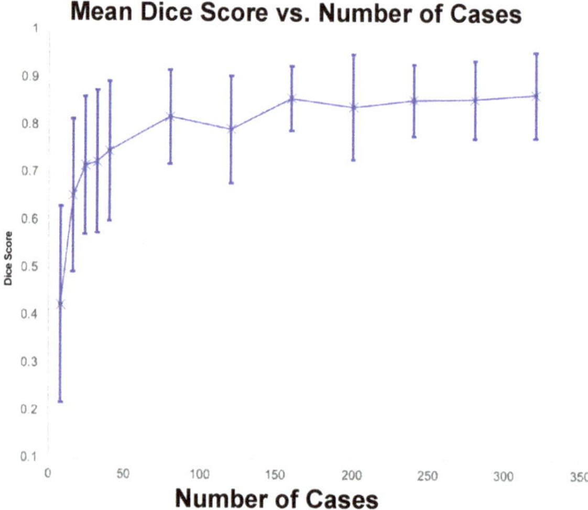

Figure 3. Dice score improved the most between 8 cases and 16 cases (0.424 to 0.653). The Dice score started to plateau after 160 cases which had a performance of 0.858. The Dice score only improved by 0.09 from 160 cases to 320 cases. The Dice score was plotted with error bars that show the standard deviation above and below that run's mean Dice score. The standard deviation was lowest at 0.076 with 240 cases and highest at 0.206 with 8 cases.

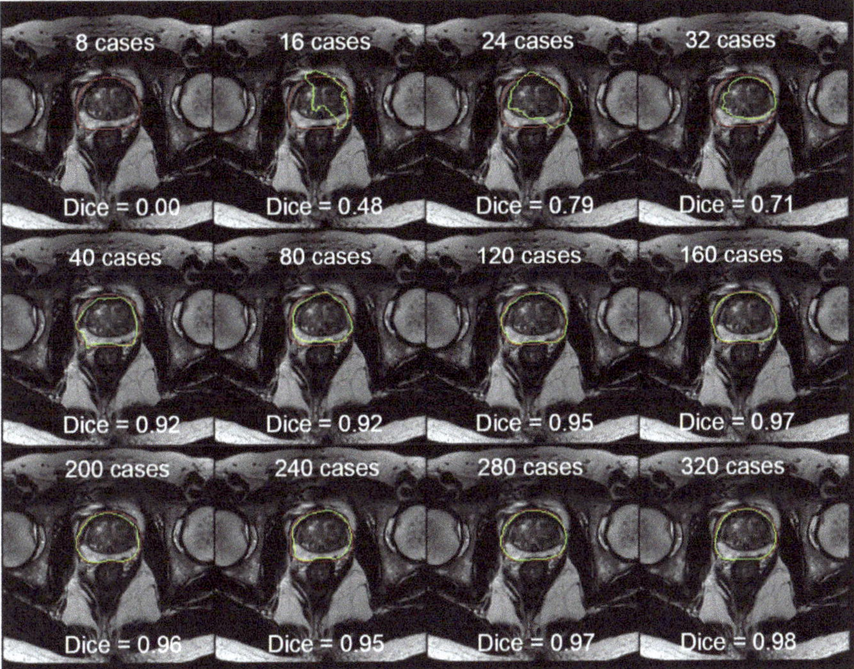

Figure 4. The performance of the U-Net was plotted for one axial slice on a single case across the different training sizes. The red line is the ground truth and the green line is the U-Net. The Dice score for one axial slice is shown in each square. The Dice score started to stabilize once the neural network trained with 160 cases.

3.2. Convolutional Neural Network Details/Statistics

The stability of the U-Net in training was evaluated (Figure 5). During training, the neural network runs that used training sizes between 8 and 40 did not converge quickly. By contrast, the neural network runs that used training sizes between 80 and 320 did converge quickly. The highest variance was 0.046 for the run that used 40 cases and the lowest variance was 0.003 for that run that used 200 cases. The training process required approximately 7 h of training time for each run. During inference, the U-Net took an average of 0.24 s per case on one GPU to complete inference.

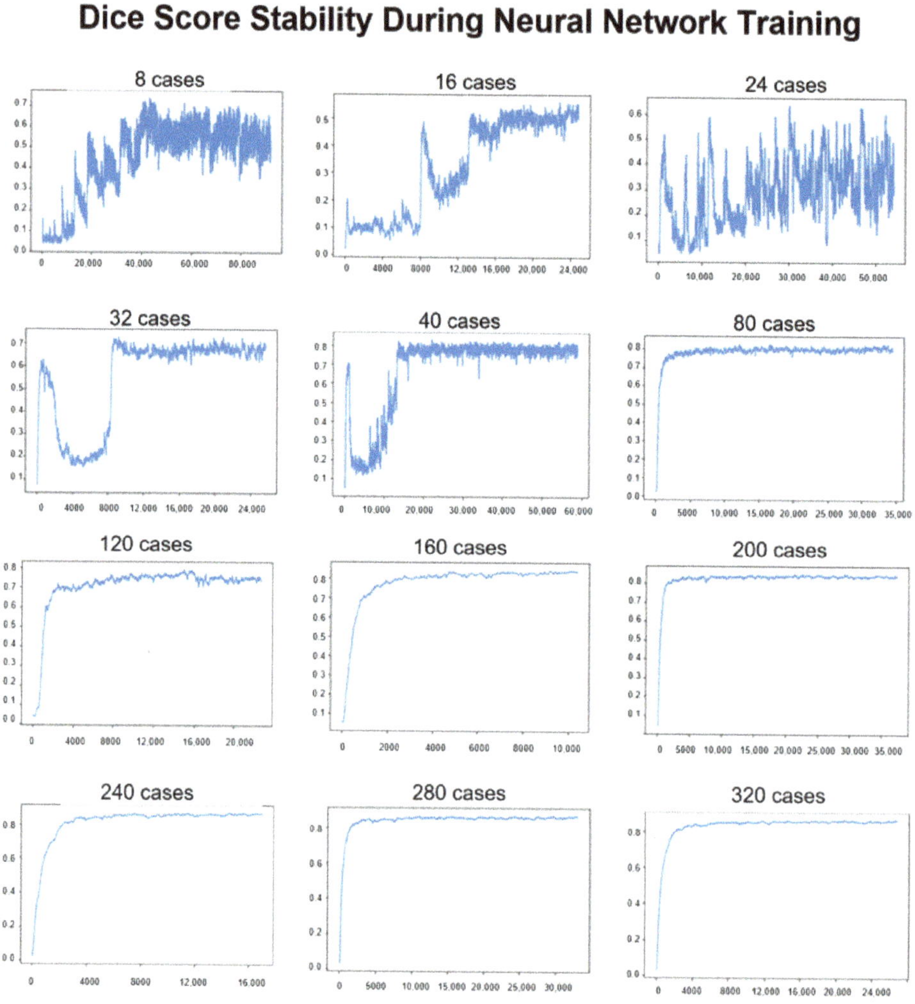

Figure 5. The number of iterations is plotted on the x-axis and the Dice score during training is plotted on the y-axis. The mean Dice score was plotted during training for the 12 different dataset sizes. The Dice score exhibited instability when training on case sizes of 8, 16, 24, 32, and 40. The Dice score stabilized more easily on case sizes of 80, 120, 160, 200, 240, 280, and 320. The Dice score variance was calculated during training; the run with 40 cases had the highest variance of 0.046 and the run with 200 cases had the lowest variance of 0.003.

4. Discussion

The purpose of this study was to explore the relationship between training size and CNN performance for prostate organ segmentation. As expected, the CNN performance plateaued with more data after 160 cases, providing a minimal increase in the Dice score. The Dice score was 0.858 at 160 cases and improved to 0.867 at 320 cases. These results confirm our hypothesis that providing more data after a certain size would only provide marginal benefits. The Dice score performance was best modeled with an asymptotic function (Equation (3)) that will converge as the number of cases increases. By using this asymptotic function (Equation (3)) for prediction, the Dice score would reach 0.871 with 500 cases and 0.877 with 1000 cases. The results also demonstrated that the selection of

U-Net as the CNN was apt due to effective prostate segmentation. U-Net's design that classifies each voxel after contraction and expansion are completed to extract unique features make it an apt network for medical imaging analysis. Since manual prostate segmentation is a tedious task [21] and took between 3 and 7 min per case for our radiologists, it is beneficial to know that more cases will not automatically translate into superior results.

Our study is unique because of its dataset size, which enabled us to find an optimal number of cases for training. In ten previous studies that also completed prostate segmentation, the dataset sizes ranged from 21 to 163 cases [22–31]. Three of these studies by Zhu et al. [28], Zhu et al. [27], and Clark et al. [26] were most comparable to our study because they also used a U-Net for their CNN. These three studies obtained Dice scores of 0.89, 0.93, and 0.89 with dataset sizes of 134, 163, and 81 cases, respectively. Although these studies did not compare training with multiple dataset sizes, their results support our findings that U-Net can achieve accurate results for prostate segmentation with a limited dataset.

Along with prostate segmentation, U-Net has demonstrated that it can segment other organs with small dataset sizes. The kidneys were accurately segmented by a U-Net in a study by Jackson et al. [32] with 89 cases. Jackson's study achieved Dice scores of 0.91 and 0.86 for the left and right kidneys, respectively [32]. Multiple U-Nets were combined together to segment multiple organs simultaneously on thorax computed tomography (CT) images in a study by Dong et al. [33]. In Dong's study, the network trained with 40 cases to obtain Dice scores of 0.97, 0.97, 0.90, and 0.87 for the left lung, right lung, spinal cord, and heart, respectively [33]. These studies demonstrate that a U-Net is a well-suited CNN for organ segmentation because of its ability to provide accurate results on small datasets. If these studies were to increase their number of cases, their Dice scores would probably improve and eventually plateau as well.

Several limitations should be considered in our study. All training data were gathered from one academic institution and two manufacturers' MRI scanners. All acquisitions were performed at 3 tesla (3T) MRI field strength and without endorectal coil. Although our CNN works well on our dataset, its ability to generalize with more prostate MRIs outside of our institution could be tested with studies from other institutions. Further work should explore the minimum amount of data for other tasks that build upon prostate organ segmentation. Different dataset sizes could be used to train networks that identify different prostate zones [34] and detect prostate lesions [35]. Along with the prostate, the training dataset size could be varied for other abdominal organs such as the kidney. These studies would serve as useful reference points for future studies that seek to optimize their neural networks. Additional work in this dataset should progress beyond prostate segmentation and detect prostate lesions. Lesion identification is a much more challenging task for AI and data augmentation with a generative adversarial network (GAN) [36] could be very useful since this technical problem lacks sufficient training data [37].

Given the popularity of AI to complete medical imaging projects that perform organ and lesion detection [38], we predict that segmentation projects will likely see diminishing returns in network performance after a threshold number of data points. As such, large datasets may not be a requirement to performing quality AI imaging research. Study teams can start with smaller datasets and evaluate performance analysis on subsets of the training data to predict the plateau effect in their datasets.

5. Conclusions

The required number of annotated cases for accurate organ segmentation with a deep learning network may be lower than expected. The marginal benefit of more data may diminish after reaching a threshold number of cases in a deep learning network. In this study of prostate organ segmentation, the U-Net CNN plateaued at 160 cases.

Author Contributions: Conceptualization, M.B., D.C., and P.C.; methodology, M.B., and R.H.; software, M.B., C.C., and P.C.; validation, M.B., C.C., and P.C.; formal analysis, M.B., C.C., and P.C.; investigation, M.B., R.H., D.C., and P.C.; resources, D.C., E.U., and P.C.; data curation, M.B., R.H., A.U., J.G.-B., E.U., and P.C.; writing—original draft preparation, M.B.; writing—review and editing, M.B., R.H., A.U., J.G.-B., M.S., and D.C.; visualization, M.B., R.H., C.C., D.C., and P.C.; supervision, R.H., D.C., and P.C.; project administration, R.H., D.C., and P.C.; funding acquisition, M.B., D.C., and P.C. All authors have read and agreed to the published version of the manuscript.

Funding: This research is supported by a Radiological Society of North America Medical Student Research Grant (RMS1902) and additionally by an Alpha Omega Alpha Carolyn L. Kuckein Student Research Fellowship.

Disclosures: Author Peter Chang, MD, is a cofounder and shareholder of Avicenna.ai, a medical imaging startup. Author Daniel Chow, MD, is a shareholder of Avicenna.ai, a medical imaging startup.

Conflicts of Interest: The authors declare no conflicts of interest. The funders had no role in the design of the study; in the collection, analyses, or interpretation of data; in the writing of the manuscript, or in the decision to publish the results.

Appendix A

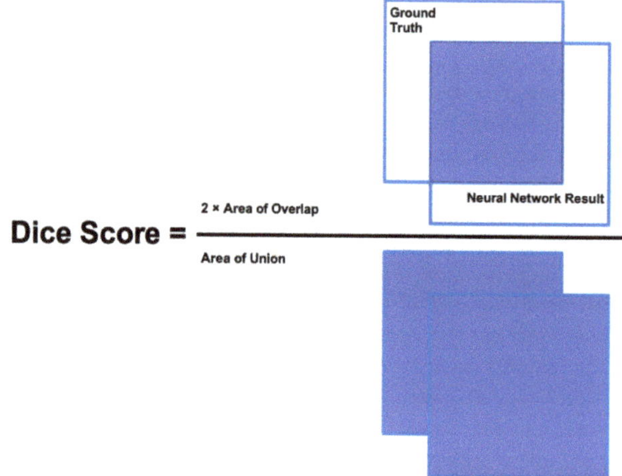

Figure A1. The Dice score is used to measure the performance of the neural network. Its range is from 0 (worst) to 1 (best). A score of 1 demonstrates perfect overlap between the ground truth and the neural network's output. A score of 0 shows that the ground truth and neural network's output have no intersection.

References

1. Litjens, G.; Kooi, T.; Bejnordi, B.E.; Setio, A.A.A.; Ciompi, F.; Ghafoorian, M.; Van Der Laak, J.A.; Van Ginneken, B.; Sánchez, C.I. A survey on deep learning in medical image analysis. *Med. Image Anal.* **2017**, *42*, 60–88. [CrossRef] [PubMed]
2. He, K.; Zhang, X.; Ren, S.; Sun, J. Deep residual learning for image recognition. In Proceedings of the IEEE Conference on Computer Vision and Pattern Recognition, Las Vegas, NV, USA, 27–30 June 2016; pp. 770–778.
3. Krizhevsky, A.; Sutskever, I.; Hinton, G.E. Imagenet classification with deep convolutional neural networks. *Adv. Neural Inf. Process. Syst.* **2012**, *25*, 1097–1105. [CrossRef]
4. Chang, P.; Kuoy, E.; Grinband, J.; Weinberg, B.; Thompson, M.; Homo, R.; Chen, J.; Abcede, H.; Shafie, M.; Sugrue, L. Hybrid 3D/2D convolutional neural network for hemorrhage evaluation on head CT. *Am. J. Neuroradiol.* **2018**, *39*, 1609–1616. [CrossRef] [PubMed]
5. Chang, P.; Grinband, J.; Weinberg, B.; Bardis, M.; Khy, M.; Cadena, G.; Su, M.-Y.; Cha, S.; Filippi, C.; Bota, D. Deep-Learning Convolutional Neural Networks Accurately Classify Genetic Mutations in Gliomas. *Am. J. Neuroradiol.* **2018**, *39*, 1201–1207. [CrossRef] [PubMed]
6. Figueroa, R.L.; Zeng-Treitler, Q.; Kandula, S.; Ngo, L.H. Predicting sample size required for classification performance. *BMC Med. Inform. Decis. Mak.* **2012**, *12*, 8. [CrossRef]

7. Fukunaga, K.; Hayes, R.R. Effects of sample size in classifier design. *IEEE Trans. Pattern Anal. Mach. Intell.* **1989**, *11*, 873–885. [CrossRef]
8. Gheisari, M.; Wang, G.; Bhuiyan, M.Z.A. A survey on deep learning in big data. In Proceedings of the 2017 IEEE International Conference on Computational Science and Engineering (CSE) and IEEE International Conference on Embedded and Ubiquitous Computing (EUC), Guangzhou, China, 21–24 July 2017; pp. 173–180.
9. Zhang, Q.; Yang, L.T.; Chen, Z.; Li, P. A survey on deep learning for big data. *Inf. Fusion* **2018**, *42*, 146–157. [CrossRef]
10. Hulsen, T.; Jamuar, S.S.; Moody, A.R.; Karnes, J.H.; Varga, O.; Hedensted, S.; Spreafico, R.; Hafler, D.A.; McKinney, E.F. From Big Data to Precision Medicine. *Front. Med. (Lausanne)* **2019**, *6*, 34. [CrossRef]
11. Kohli, M.D.; Summers, R.M.; Geis, J.R. Medical Image Data and Datasets in the Era of Machine Learning-Whitepaper from the 2016 C-MIMI Meeting Dataset Session. *J. Digit. Imaging* **2017**, *30*, 392–399. [CrossRef]
12. Baltres, A.; Al Masry, Z.; Zemouri, R.; Valmary-Degano, S.; Arnould, L.; Zerhouni, N.; Devalland, C. Prediction of Oncotype DX recurrence score using deep multi-layer perceptrons in estrogen receptor-positive, HER2-negative breast cancer. *Breast Cancer (Tokyo Jpn.)* **2020**. [CrossRef]
13. Zemouri, R.; Omri, N.; Morello, B.; Devalland, C.; Arnould, L.; Zerhouni, N.; Fnaiech, F. Constructive deep neural network for breast cancer diagnosis. *IFAC-PapersOnLine* **2018**, *51*, 98–103. [CrossRef]
14. Ker, J.; Wang, L.; Rao, J.; Lim, T. Deep learning applications in medical image analysis. *IEEE Access* **2018**, *6*, 9375–9389. [CrossRef]
15. Cho, J.; Lee, K.; Shin, E.; Choy, G.; Do, S. How much data is needed to train a medical image deep learning system to achieve necessary high accuracy? *arXiv* **2015**, arXiv:1511.06348.
16. Lakhani, P. Deep convolutional neural networks for endotracheal tube position and X-ray image classification: Challenges and opportunities. *J. Digit. Imaging* **2017**, *30*, 460–468. [CrossRef] [PubMed]
17. Ronneberger, O.; Fischer, P.; Brox, T. U-Net: Convolutional Networks for Biomedical Image Segmentation. In Proceedings of the International Conference on Medical Image Computing and Computer-Assisted Intervention, Munich, Germany, 5–9 October 2015; pp. 234–241.
18. Kingma, D.P.; Ba, J. Adam: A method for stochastic optimization. *arXiv* **2014**, arXiv:1412.6980.
19. Zou, K.H.; Warfield, S.K.; Bharatha, A.; Tempany, C.M.; Kaus, M.R.; Haker, S.J.; Wells III, W.M.; Jolesz, F.A.; Kikinis, R. Statistical validation of image segmentation quality based on a spatial overlap index1: Scientific reports. *Acad. Radiol.* **2004**, *11*, 178–189. [CrossRef]
20. Virtanen, P.; Gommers, R.; Oliphant, T.E.; Haberland, M.; Reddy, T.; Cournapeau, D.; Burovski, E.; Peterson, P.; Weckesser, W.; Bright, J. SciPy 1.0: Fundamental algorithms for scientific computing in Python. *Nat. Methods* **2020**, *17*, 261–272. [CrossRef] [PubMed]
21. Clark, T.; Zhang, J.; Baig, S.; Wong, A.; Haider, M.A.; Khalvati, F. Fully automated segmentation of prostate whole gland and transition zone in diffusion-weighted MRI using convolutional neural networks. *J. Med. Imaging (Bellingham)* **2017**, *4*, 041307. [CrossRef]
22. Rundo, L.; Militello, C.; Russo, G.; Garufi, A.; Vitabile, S.; Gilardi, M.; Mauri, G. Automated prostate gland segmentation based on an unsupervised fuzzy C-means clustering technique using multispectral T1w and T2w MR imaging. *Information* **2017**, *8*, 49. [CrossRef]
23. Ghose, S.; Mitra, J.; Oliver, A.; Marti, R.; Lladó, X.; Freixenet, J.; Vilanova, J.C.; Sidibé, D.; Meriaudeau, F. A random forest based classification approach to prostate segmentation in MRI. *Miccai Grand Chall. Prostate MR Image Segm.* **2012**, *2012*, 125–128.
24. Tian, Z.; Liu, L.; Zhang, Z.; Fei, B. PSNet: Prostate segmentation on MRI based on a convolutional neural network. *J. Med. Imaging* **2018**, *5*, 021208. [CrossRef] [PubMed]
25. Karimi, D.; Samei, G.; Kesch, C.; Nir, G.; Salcudean, S.E. Prostate segmentation in MRI using a convolutional neural network architecture and training strategy based on statistical shape models. *Int. J. Comput. Assist. Radiol. Surg.* **2018**, *13*, 1211–1219. [CrossRef] [PubMed]
26. Clark, T.; Wong, A.; Haider, M.A.; Khalvati, F. Fully Deep Convolutional Neural Networks for Segmentation of the Prostate Gland in Diffusion-Weighted MR Images. In Proceedings of the International Conference Image Analysis and Recognition, Montreal, QC, Canada, 5–7 July 2017; pp. 97–104.
27. Zhu, Y.; Wei, R.; Gao, G.; Ding, L.; Zhang, X.; Wang, X.; Zhang, J. Fully automatic segmentation on prostate MR images based on cascaded fully convolution network. *J. Magn. Reson. Imaging* **2019**, *49*, 1149–1156. [CrossRef]

28. Zhu, Q.; Du, B.; Turkbey, B.; Choyke, P.L.; Yan, P. Deeply-supervised CNN for prostate segmentation. In Proceedings of the 2017 International Joint Conference on Neural Networks (Ijcnn), Anchorage, Alaska, 14–19 May 2017; pp. 178–184.
29. Milletari, F.; Navab, N.; Ahmadi, S.-A. V-net: Fully convolutional neural networks for volumetric medical image segmentation. In Proceedings of the 2016 Fourth International Conference on 3D Vision (3DV), Stanford, CA, USA, 25–28 October 2016; pp. 565–571.
30. Wang, B.; Lei, Y.; Tian, S.; Wang, T.; Liu, Y.; Patel, P.; Jani, A.B.; Mao, H.; Curran, W.J.; Liu, T. Deeply supervised 3D fully convolutional networks with group dilated convolution for automatic MRI prostate segmentation. *Med. Phys.* **2019**, *46*, 1707–1718. [CrossRef] [PubMed]
31. Cheng, R.; Roth, H.R.; Lu, L.; Wang, S.; Turkbey, B.; Gandler, W.; McCreedy, E.S.; Agarwal, H.K.; Choyke, P.; Summers, R.M. Active appearance model and deep learning for more accurate prostate segmentation on MRI. In Proceedings of the Medical Imaging 2016: Image Processing, San Diego, CA, USA, 27 February 2016; p. 97842I.
32. Jackson, P.; Hardcastle, N.; Dawe, N.; Kron, T.; Hofman, M.; Hicks, R.J. Deep learning renal segmentation for fully automated radiation dose estimation in unsealed source therapy. *Front. Oncol.* **2018**, *8*, 215. [CrossRef] [PubMed]
33. Dong, X.; Lei, Y.; Wang, T.; Thomas, M.; Tang, L.; Curran, W.J.; Liu, T.; Yang, X. Automatic multiorgan segmentation in thorax CT images using U-net-GAN. *Med. Phys.* **2019**, *46*, 2157–2168. [CrossRef] [PubMed]
34. Toth, R.; Ribault, J.; Gentile, J.; Sperling, D.; Madabhushi, A. Simultaneous Segmentation of Prostatic Zones Using Active Appearance Models With Multiple Coupled Levelsets. *Comput. Vis. Image Underst.* **2013**, *117*, 1051–1060. [CrossRef]
35. Lay, N.S.; Tsehay, Y.; Greer, M.D.; Turkbey, B.; Kwak, J.T.; Choyke, P.L.; Pinto, P.; Wood, B.J.; Summers, R.M. Detection of prostate cancer in multiparametric MRI using random forest with instance weighting. *J. Med. Imaging* **2017**, *4*, 024506. [CrossRef]
36. Mao, X.; Li, Q.; Xie, H.; Lau, R.Y.; Wang, Z.; Smolley, S.P. Least squares generative adversarial networks. In Proceedings of the 2017 IEEE International Conference on Computer vision (ICCV), Venice, Italy, 22–29 October 2017; pp. 2794–2802.
37. Bardis, M.D.; Houshyar, R.; Chang, P.D.; Ushinsky, A.; Glavis-Bloom, J.; Chahine, C.; Bui, T.-L.; Rupasinghe, M.; Filippi, C.G.; Chow, D.S. Applications of Artificial Intelligence to Prostate Multiparametric MRI (mpMRI): Current and Emerging Trends. *Cancers* **2020**, *12*, 1204. [CrossRef]
38. Zemouri, R.; Zerhouni, N.; Racoceanu, D. Deep learning in the biomedical applications: Recent and future status. *Appl. Sci.* **2019**, *9*, 1526. [CrossRef]

Article

Novel Biased Normalized Cuts Approach for the Automatic Segmentation of the Conjunctiva

Giovanni Dimauro [1,*] and Lorenzo Simone [2]

1 Department of Computer Science, University of Bari, 70125 Bari, Italy;
2 Department of Computer Science, University of Pisa, 56127 Pisa, Italy; l.simone3@studenti.unipi.it
* Correspondence: giovanni.dimauro@uniba.it

Received: 18 May 2020; Accepted: 9 June 2020; Published: 14 June 2020

Abstract: Anemia is a common public health disease diffused worldwide. In many cases it affects the daily lives of patients needing medical assistance and continuous monitoring. Medical literature states empirical evidence of a correlation between conjunctival pallor on physical examinations and its association with anemia diagnosis. Although humans exhibit a natural expertise in pattern recognition and associative skills based on hue properties, the variance of estimates is high, requiring blood sampling even for monitoring. To design automatic systems for the objective evaluation of pallor utilizing digital images of the conjunctiva, it is necessary to obtain reliable automatic segmentation of the eyelid conjunctiva. In this study, we propose a graph partitioning segmentation approach. The semantic segmentation procedure of a diagnostically meaningful region of interest has been proposed for exploiting normalized cuts for perceptual grouping, thereby introducing a bias towards spectrophotometry features of hemoglobin. The reliability of the identification of the region of interest is demonstrated both with standard metrics and by measuring the correlation between the color of the ROI and the hemoglobin level based on 94 samples distributed in relation to age, sex and hemoglobin concentration. The region of interest automatically segmented is suitable for diagnostic procedures based on quantitative hemoglobin estimation of exposed tissues of the conjunctiva.

Keywords: semantic segmentation; pattern recognition; hemoglobin; anemia; human tissues; conjunctiva; non-invasive medical device

---

1. Introduction

1.1. Background

Anemia is a blood disorder in which the number of red blood cells is inadequate to carry oxygen to human tissues and organs. It affects about a third of the global population, being the most common blood disorder according to the epidemiological results [1–3]. Each different form of this condition has its specific underlying causes. The process of erythrocyte production in the blood involves bone marrow and erythropoietin, a hormone produced by the kidneys, which regulates the process of erythropoiesis, favoring a constant rate of change in the number of erythrocytes in the blood [4]. Adequate production of red blood cells prevents conditions such as anemia and tissue hypoxia. To promote normal erythropoiesis, correct hemoglobin synthesis is required. Hemoglobin, the iron-containing protein, represents the predominant protein found in erythrocytes, responsible for transporting oxygen from the lungs to the other tissues. Anemia caused by deficiencies of the aforementioned factors results in production patterns of abnormal and different erythrocytes [5]. Diagnosing anemia requires in most cases a complete blood count (CBC) to check different properties, including hemoglobin and hematocrit levels. Each physiological need depends on several factors, such as gender, age, different stages of pregnancy and altitude. The thresholds presented in Table 1 are used to diagnose anemia in individuals in a screening or clinical setting according to World Health Organization diagnostic guidelines [6].

Table 1. Hemoglobin (Hb) thresholds used to define anemia living at sea level according to the World Health Organization guidelines [6].

Age Group	No Anemia	Mild Anemia	Moderate Anemia	Severe Anemia
Children 5–11 years	$\geq$ 11.5 g/dL	11–11.4 g/dL	8–10.9 g/dL	<8 g/dL
Children 12–14 years	$\geq$ 12 g/dL	11–11.9 g/dL	8–10.9 g/dL	<8 g/dL
Non-pregnant women	$\geq$ 12 g/dL	11–11.9 g/dL	8–10.9 g/dL	<8 g/dL
Pregnant women	$\geq$ 11 g/dL	10–10.9 g/dL	7–9.9 g/dL	<7 g/dL
Men	$\geq$ 13 g/dL	11–12.9 g/dL	8–10.9 g/dL	<8 g/dL

There has always been a worldwide interest in providing simple, cheap and robust procedures to measure hemoglobin without requiring specialized primary health-care workers or medical laboratories [7]. In response to this need, WHO developed the hemoglobin color scale (HCS) in 2001. It consists of a small card of six shades of red from lighter to darker representing a hemoglobin g/dL concentration from 4 to 14 with a step size of 2 g/dL. The specificity of this method has been disputed in literature; for instance, in 2005 14 studies mostly reported a high sensitivity for detecting anemia (75–97%) [8]. Nevertheless, what is crucial about HCS is its potential for opening the way to different approaches requiring a mixture of expertise from different disciplines, such as computer science, in the future. Like other diagnostic-clinical and analytical-laboratory medical disciplines that are beginning to make extensive use of image, sound or signal analysis; and machine and deep learning techniques [9–18], it is worthwhile to invest in research and development of technologies such as those we deal with in this paper, with the dual purpose of significantly reducing the costs borne by the national health systems and powering the healthcare and medical services that would be exempted from a considerable amount of practically useless activities. Since the importance of the objective evaluation of the pallor of the conjunctiva has been understood, a lot has been done. Numerous researchers have worked to develop methods, techniques and devices to make the estimate of the level of hemoglobin or the determination of the condition of severe anemia, in a non-invasive way, as reliable as possible. We will report a summary of this path in the section "Related Works."

1.2. Haemoglobin Spectrophotometry

HCS and physical examination of exposed tissues such as palpebral conjunctiva or nail beds both rely on how humans perceive colors related to the optical spectrum [19]. To better analyze and handle this phenomenon from a computer vision point of view, a chemical insight is required. Spectrophotometry in chemistry is defined as quantitative measurements of the reflective or absorption properties of a material from a wavelength perspective. The spectra of the hemoglobin molecule vary based on whether it is bound to oxygen, carbon monoxide or nothing; the the latter is also called deoxygenated Hb [20].

We relied on experimental literature data [21] for the absorption spectra of hemoglobin used for both plots in Figure 1. The absorption coefficient μ_a^{Hb} for HbO_2 and Hb is calculated as follows:

$$\mu_a^{Hb}(\lambda) = \frac{2.303 \times e_{Hb}(\lambda)[\frac{L}{cm \times mol}] \times 150[g/L]}{M_{Hb}[g/mol]}, \tag{1}$$

where $e_{Hb}(\lambda)[\frac{L}{cm \times mol}]$ is the Hb molar extinction coefficient and $M_{Hb}[g/mol]$ is the Hb gram molecular weight, assuming a concentration of 150 grams per liter.

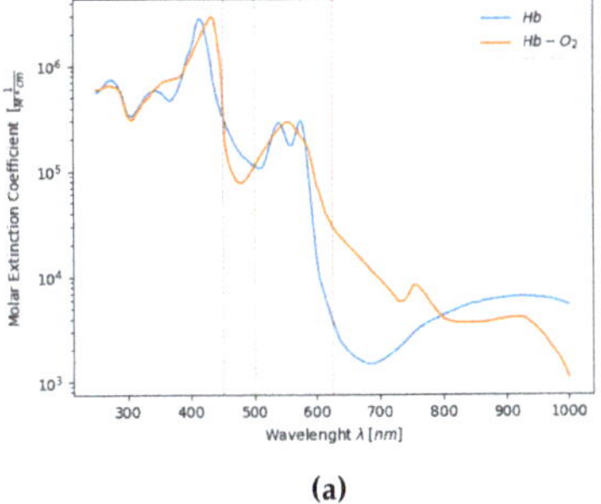

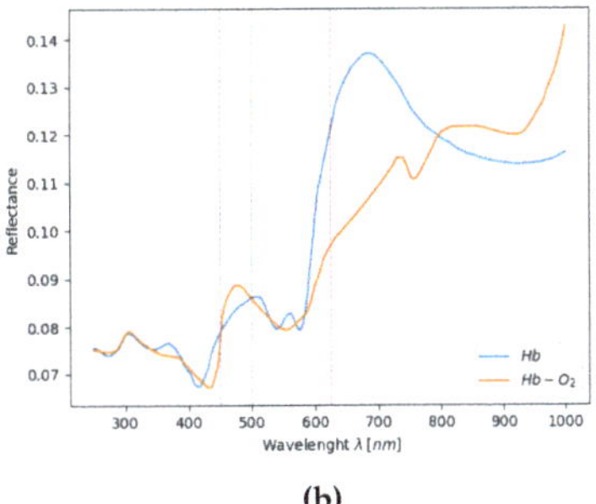

(a) (b)

Figure 1. Plots visualizing optical absorption and reflectance of Hb and HbO_2, vertical dashed lines are related to human perception of colors associated with (λ). (**a**) Molar extinction coefficient (ϵ) related to absorbance over wavelength (λ) considering 15 g/dL of hemoglobin concentration and 1 cm cuvette. (**b**) Derived reflectance plot of absorbance under same constants.

Over the years, the palpebral conjunctiva has been a good spot to diagnose anemia, representing a highly vascular area characterized by several capillaries. In [22] a multi-layered tissue model is proposed and investigated to approximate the lower eyelid with seven layers: conjunctival epithelium, tarsal plate, orbicularis oculi, subcutaneous tissue, dermis, epidermis and stratum corneum on the outside of the eyelid tissue. The conjunctiva is perfused from the ascending branch of the posterior conjunctival artery. The presence of interweaving capillary networks penetrating several layers of the model, with the mucous membrane being highly transparent, allows for model approximations for the digital image domain. As already visually described by Figure 1, Hb and HbO_2 both absorb wavelengths from 275 to about 550 nm corresponding to a visible spectrum from purple to light green. Each frequency above 600 nm is highly reflected, matching with colors from orange to dark red. A typical human eye is known to be aware of wavelengths in a range from 380 to 740 nm. The cytoplasm of the red blood cell is rich in hemoglobin, that being responsible for the reddish appearance of exposed tissues and blood in general. Laboratory-based experiments conducted in [23,24], inspired us to start from those results to accomplish segmentation and digital image analysis related to hemoglobin.

1.3. Related Works

Over the years many researchers have put in effort toward developing non-invasive methods for anemia detection through hemoglobin estimation. The relevance of conjunctiva hue in the clinical evaluation of anemia was tested in [25] for 219 healthy ambulatory subjects. Three educated non-clinicians, appropriately trained, overall agreed on conjunctiva hue performing with kappa coefficients between 0.27 and 0.34. As a result, hue variation strictly depends on the objective of the assessment and training of field personnel. Comparing earlier results obtained by physical examination and the latest digital photography, the latter is minimizing variance, optimizing specificity and sensitivity by using machine learning and automatic segmentation procedures. Establishing the most successful technology still leaves questions about the best region to analyze exploiting color properties associated with better results. Studies in [26] from an ophthalmology point of view open a debate for correlation of anemia between bulbar conjunctival blood column and palpebral conjunctival hue (PCH). From the results of this study, it seems that the bulbar conjunctiva can be successfully included in the set of interesting features, achieving slightly less specificity than PCH, but higher sensitivity. Paradigms of non-invasive and on-demand diagnostics based on smartphone and digital images are spreading due to the advancing of remote diagnosis and affordability [27–29]. A smartphone camera-based application monitoring blood hemoglobin concentration has been developed in [30]. Utilizing a light source pointed to the patient's finger, they performed a chromatic analysis on 31 samples, achieving sensitivity and precision of 85.7% and 76.5% respectively; they received Food and Drug Administration agreement. Another smartphone-based self-screening tool is depicted in [31] utilizing fingernail beds digital images. Patients select the regions of interest by themselves, corresponding to the nailbeds, and a result is then displayed on the smartphone screen; camera flash reflections and white spots which may affect Hgb level measurements are removed with a quality control algorithm. They reported an accuracy of ± 0.92 g/dL^{-1} of CBC hemoglobin level with personalized calibration, suggesting the relevance of those systems as a monitoring utility. In our study, we analyzed assumptions from related past works and the clinical correlation between conjunctival pallor and anemia condition [32], proposing a fully automated segmentation algorithm. Throughout this process, color features from hemoglobin reflectance spectrum provide a key role in biasing towards a region of interest proposal.

In the literature, few works deal with the automatic segmentation of the conjunctiva. In particular, reference [33] proposes a method for the automatic segmentation of the palpebral conjunctiva that carries out an image processing process based on the equalization of the image in RGB, filter unsharp masking and red channel masking. In [34] the authors developed an algorithm for automatically segmenting the image by finding a "distinctly red" region, bounded by two parallel long-running edges at the top and the bottom; this is achieved by combining the Canny edge detection technique with morphological operations in the CIELAB color space. However, with the aim of estimating anemia, they stated that their method of segmenting was less reliable than manual conjunctiva segmentation made by an expert physician. In [35] the authors use a threshold triangle (which uses triangle algorithm for thresholding) for binary differentiation between the palpebral conjunctiva and background.

1.4. Image Capturing Methodology

The technique adopted to capture digital images of a patient's conjunctiva was based on the latest approach of a research study conducted in [36–38]. As a recap, the main requirements to designing an effective tool for estimating the condition of anemia through the use of digital images of the palpebral conjunctiva would be:

- Provide an easy to us;e device with affordable hardware components
- Its usage should not require trained medical personnel;
- It should provide remote diagnosis and telemedicine conveniences.

The acquisition system is shown in Figure 2. It consists of a macro-lens assembled into a specially designed, 3D-printed lightened spacer Figure 2a and a typical smartphone as in the real-life application Figure 2b. The lens can take high-resolution images being attached to a smartphone (we used the Aukey PL-M1 25 mm 10x macro lens). The LED lights can be powered directly from the smartphone or a battery applied to the cover of a smartphone. The lens is fixed on the plastic cover of the smartphone: this device allows for obtaining high resolution images close to the eye, insensitive to the ambient lighting conditions.

The dataset used in the present study, which will be described later, has been created with a Samsung S6 smartphone.

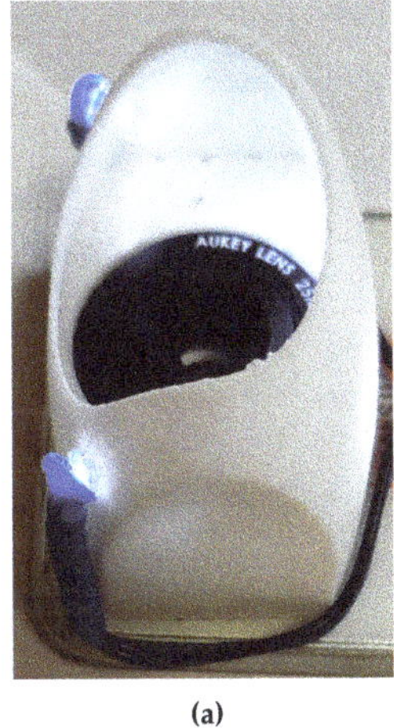

(a)

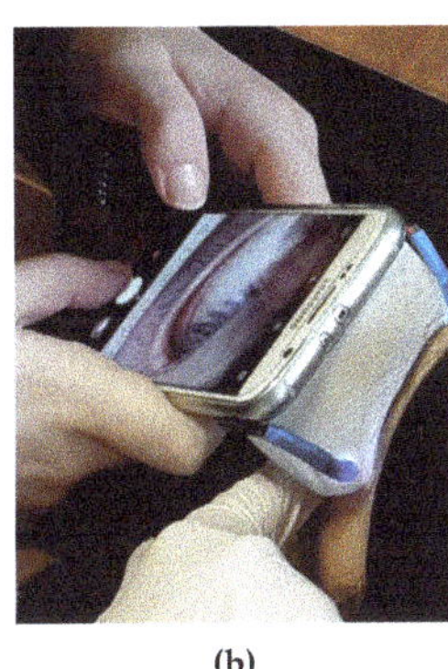
(b)

Figure 2. (**a**) The acquisition device consists of a special spacer and a macro lens to acquire images with a high-resolution smartphone at close range; (**b**) the moment of the acquisition of an image of the conjunctiva.

2. Proposed Method

Each digital image from the dataset is converted into an RGB color space matrix representation. The segmentation process can be summarized in three different phases: dimensionality reduction by clustering approach, grouping as graph partitioning and a final ROI extraction. The introduction of a preliminary clustering step determines a speed up in N-Cuts performance arising from the theoretical proofs by the N-Cuts original paper regarding computational complexity in terms of both space and time. The algorithm constructing a region adjacency graph (RAG) does not consider each pixel from the original resolution anymore, but groups of them preserving spatial and color differences amongst them. Finally, we aim at grasping a non-linear relation between brightness intensities from the red and green channels, based on previous assumptions of reflectance rate by a spectrophotometry point of view.

2.1. K-Means Dimensionality Reduction

The objective of a clustering task is grouping data instances into subsets maximizing a similarity measure, while different instances should belong to different groups [39–41]. We applied the principles from k-means clustering to image segmentation tasks. The main goal in this phase is to produce a feature space similarly to Voronoi diagrams for planes, reducing the complexity of the graph representing the original image. Each pixel from now on will be referred to as a vector in a five-dimensional space: x and y coordinates from the matrix; R, G and B channel intensities from color representation.

$$f(x,y) = \vec{p} = \alpha_x \vec{p_x} + \alpha_y \vec{p_y} + \alpha_r \vec{p_r} + \alpha_g \vec{p_g} + \alpha_b \vec{p_b} \quad (2)$$

This approach allows us to iteratively minimize the sum of distances from each pixel to its cluster centroid. We briefly summarize the steps of the algorithm as follows:

1. Initialize centroid vectors.
2. Pixels retain spatial as well as color features, allowing us to define an appropriate weighted Euclidean distance as a measure of similarity between them. For each of them, calculate the distance d between the centroid and each pixel of the image defined as:

$$d(\vec{u},\vec{v}) = \|\vec{u}-\vec{v}\| = \sqrt{(u_x-v_x)^2+(u_y-v_y)^2+(u_r-v_r)^2+(u_g-v_g)^2+(u_b-v_b)^2} \quad (3)$$

3. Each pixel is assigned to the centroid minimizing d.
4. Recalculate the position of each centroid c_k where $\overrightarrow{p_{ki}}$ is the i_{th} pixel contained in k_{th} centroid using the relation:

$$c_k = \frac{1}{n}\sum_{i=1}^{n}\overrightarrow{p_{ki}} \quad (4)$$

This approach included in the broader field of unsupervised learning approaches, consists of initial batch updates, in which at each step we reassign points to their nearest cluster centroid, followed by cluster centroid recalculations. In online updates, the points are reassigned only if reducing the sum of intra-cluster distances. Those updates already converge towards a local minimum in short order.

In Figure 3, the original image is processed with a three-dimensional (R, G and B) space and in the last picture with a five-dimensional model including both color and spatial features. In the latter, there is not an increase in computational complexity since the only calculation affected is the distance function. However, in each digital image analyzed, the intra-cluster variance is minimized efficiently with properly outlined boundaries in between each group of pixels. The classified instances closer to mucocutaneous junction are noisy in the first approach, while on the second one each semantic class (iris, pupil, sclera, eyelid, and conjunctiva) appears as a compact union of clusters.

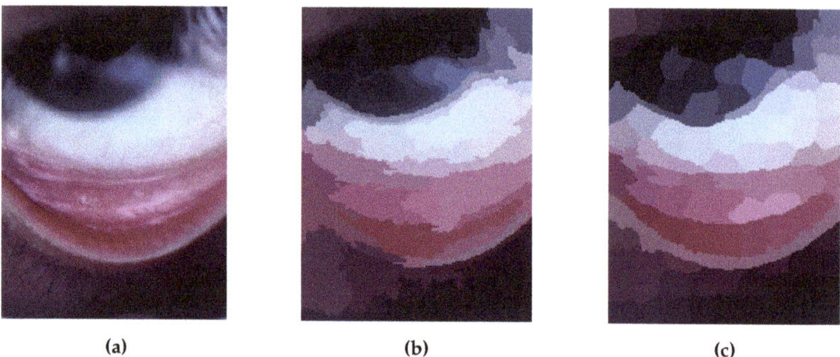

Figure 3. (**a**) Original digital image acquired; (**b**) k-means clustering procedure using only three dimensional (R,G and B) channels from color space; (**c**) proposed k-means procedure with a model in five dimensions retaining both spatial and color properties.

2.2. Normalized Cuts Segmentation

K-means as a clustering algorithm is a valuable approach for exploiting local impressions of a scene, but it lacks in providing a global or hierarchical perspective. For this reason, we take advantage of a grouping algorithm treating the segmentation task as a graph partitioning problem, such as NCuts. It has a better ability to generalize when applied to different scenarios. Conventionally, the normalized cut is an unbiased measure of dissimilarity between graph subgroups [42]. We have converted the set of superpixels from a five-dimensional feature space in a weighted undirected graph $G = (V, E)$. Each point is included in the set of nodes having one edge for each pair of vertices.

The region adjacency graph is constructed based on precomputed areas from the k-means segmentation algorithm. Each connection amongst them is depicted in Figure 4b and representable in a weight matrix W. The edge weight w_{ij} from node i to node j is defined as in the standard approach of normalized cuts as a product of a feature similarity and a spatial term. $X(i)$ is the coordinate vector of the centroid pixel and $F(i)$ is a feature vector based on averaged R, G and B intensities of each pixel in the area. The value r acts as a proximity threshold based on the Euclidean distances amongst precomputed centroids. In our specific application we have tried different configurations ranging from 3 to 100, regulating the sparsity of the weight matrix but not impacting the segmentation outcome. Weights and features are described by the following equations:

$$w_{i,j} = e^{-\frac{\|F(i)-F(j)\|_2^2}{\sigma_I}} * \begin{cases} e^{\frac{-\|X(i)-X(j)\|_2^2}{\sigma_X}}, & \text{if } \|X(i)-X(j)\|_2 < r \\ 0, & \text{otherwise} \end{cases} \tag{5}$$

$$F(i) = \left[\frac{1}{n}\sum_{j=1}^{n} p_{jr} \quad \frac{1}{n}\sum_{j=1} n p_{jg} \quad \frac{1}{n}\sum_{j=1} n p_{jb} \right] \tag{6}$$

The algorithm is capable of extracting significant components from each sample from the dataset, avoiding intra-cluster variations.

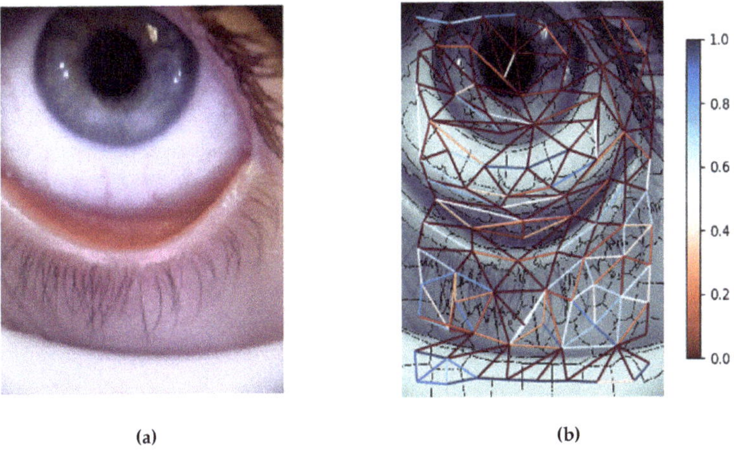

Figure 4. (**a**) Acquired sample; (**b**) region adjacency graph (RAG) displaying a measure of similarity between each region. The center of each node is considered a vertex. For each connection between two regions, there is an associated colored line according to the measure of similarity.

In Figure 5, we added a visual semantic description of the resulting cuts. With this phase, we raise the level of abstraction of the segmentation, starting from the clusters of Figure 3; we end up with features closer to an anatomical perspective. The small gap in colors between the conjunctival area and mucocutaneous junction is perfectly delineated in each sample from the dataset, paving the way for a machine-learning-based anemia estimator.

In the proposed segmentation output from Figure 5, a recursive approach could be run to further decompose regions of interest from the conjunctival area. As an example, this could lead to a better parting of the two conjunctivae, palpebral and forniceal, so as to contribute to the open debate about the prevalence of one or the other as the best estimator of anemia [43]. In fact, the palpebral conjunctiva highlights the vascularization of the underlying area better than the forniceal and probably allows highlighting minimal variations of blood color. The assumption seems confirmed by scientific literature. However, some authors take into consideration the whole conjunctiva, including both palpebral and forniceal, to construct and validate their models. It is still an open problem. Furthermore, in [43] the authors state that it should be interesting to establish whether the investigations carried out on a small portion of the conjunctiva can be sufficient and position independent. In fact, the sparsity and density of the blood micro-vessels can change in different parts of the eyelid. Therefore, the recursive identification of further clusters can help to answer the above questions.

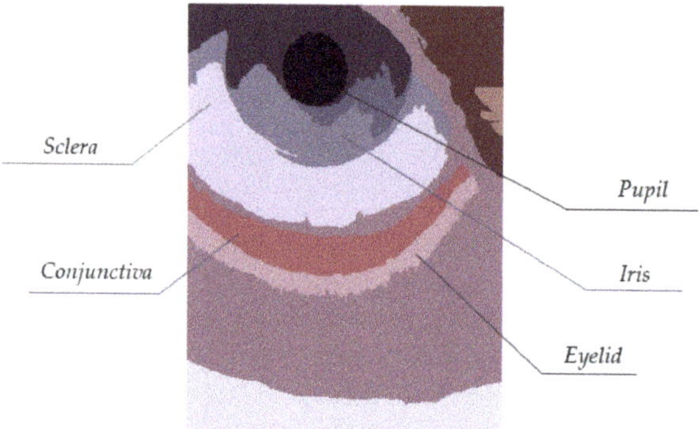

Figure 5. Segmentation output result with semantic class description of eye anatomy.

2.3. Hemoglobin Heatmap Coefficients

In medical image or radar signal processing tasks, contrast enhancement is a widely used technique in various applications, ranging from improving the quality of photographs acquired in poor conditions [44] to emphasizing regions of interest [45,46]. Histogram equalization is one of the most common approach due to its simple mechanism and effectiveness, but as a drawback, image brightness usually changes after the procedure, caused by its flattening behavior. In our study the objective is focused on approximating the spectrophotometry multi-layered reflectance model investigated in Section 1.2, grasping a mathematical description for digital images. In the literature several studies apply spectral domain scanning, resulting in a time-consuming acquisition process and expensive equipment. This approach does not fit our needs of developing a cheap, non-invasive diagnostic tool. An example of an ill-posed problem known as spectral reconstruction from an RGB scene has been conducted with deep learning techniques in [47,48]. Lastly, researches are highly promoting the validity of these approaches, but despite this, our application domain allows us to further reduce the solution required. Our method, interpreting the image as a signal, performs a pixel pointwise non-linear transformation from red and green color space values, returning a coefficient highlighting vascularized regions. In the literature, the ratio between R and G channels has often been used as a guide to spot those areas, thereby finding the highest values in forniceal and palpebral conjunctival tissues. We propose a generalized logistic function filtering technique including more flexibility than a standard sigmoid. Considering an image I as a vector in three channel functions based on grid coordinates, we obtain the following σ' transformation:

$$I(x,y) = \begin{bmatrix} r(x,y) \\ g(x,y) \\ b(x,y) \end{bmatrix}, \qquad \sigma'(I,x,y) = \frac{1}{1+e^{-\alpha\left(\frac{I_r(x,y)}{I_g(x,y)}-\beta\right)}} \tag{7}$$

The parameter α determines the slope of the function, emphasizing the discrepancy in terms of ratio between color channels; β acts as a minimum ratio threshold for the activation of each pixel.

A comparison of the behavior of standard and generalized logistic function with parameterization $\alpha = 4$ and $\beta = 2$ is depicted in Figure 6. This parameterization yielded results with a remarkable capability of generalizing well in diagnostic imaging ranging from conjunctival tissue to endoscopic domains. Increasing values of α related to the steepness, tend towards the trivial case of a binarization step function losing information about the relationship underlying a variety of brightness ratios. An application of this model is illustrated in Figure 7 useful for digital images of the conjunctival region.

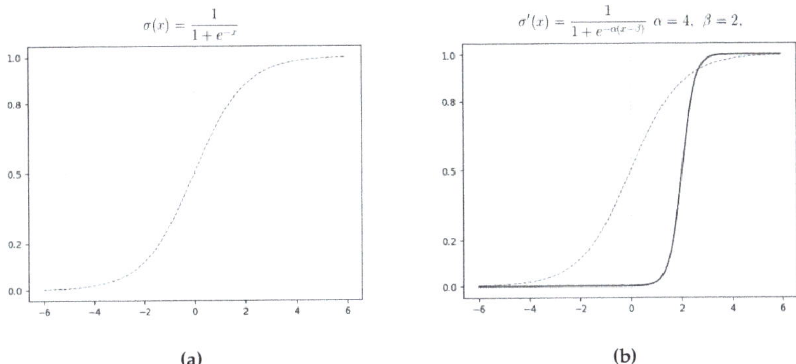

Figure 6. (**a**) Standard logistic function plot. (**b**) Generalized logistic function plot using parameters $\alpha = 4$ and $\beta = 2$.

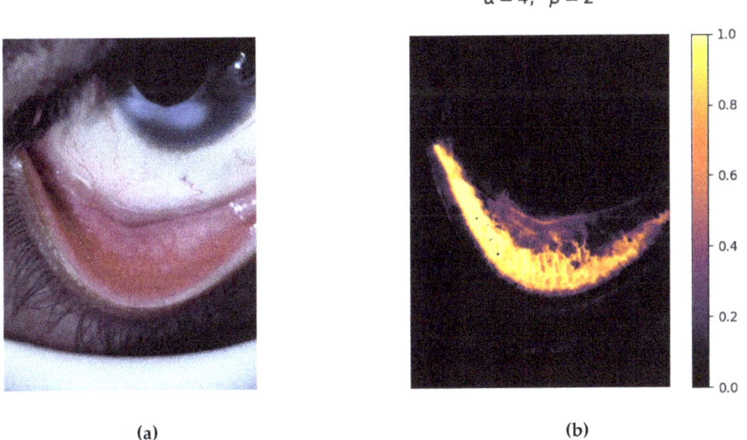

Figure 7. (**a**) Acquired sample. (**b**) Heatmap plot of the scoring matrix displaying the magnitudes of the coefficients computed by applying the generalized sigmoid function on the acquired sample.

The real values range from 0 to 1 according to σ' function definition. The filtering process produces a scoring matrix assigning lower values to the background, including the sclera, pupil, iris, eyelid and white support platform from the device. Palpebral and forniceal conjunctiva are primarily perfused by both internal and external carotid arteries; this is reflected in high values from the scoring matrix ranging from 0.7 to 1, and the respective blood vessels are significantly highlighted, as shown in Figure 7b.

Since we are interested in obtaining a semantic interpretation out of the regions proposed by NCut, the matrix of coefficients acts as an effective bias for calculating the probability distribution of each class. Edge weights crossed by aggregated pixels resulting from σ' are strengthened or decreased, resulting in a region proposal based on the magnitude of the connection. In Figure 8, we provide a subset of 10 digital images from the dataset, showing the qualitative difference between the proposed semantic segmentation (top row) and the manually segmented ground truth (second row). In Figure 9 we provide two samples of erroneous acquisitions in order to show the robustness of the proposed segmentation in unusual conditions; in fact, only images with excellent characteristics can provide useful information for the correct estimation of anemia.

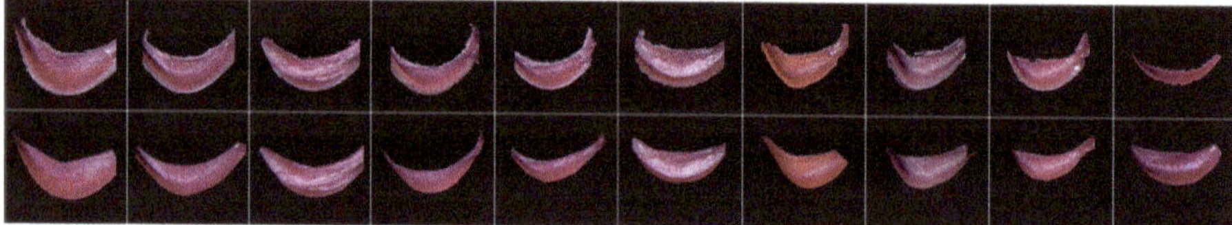

Figure 8. The top row represents a subset of samples automatically segmented with the proposed approach. The ordered second row depicts the mapping with the manual segmentation ground truth of the conjunctival region.

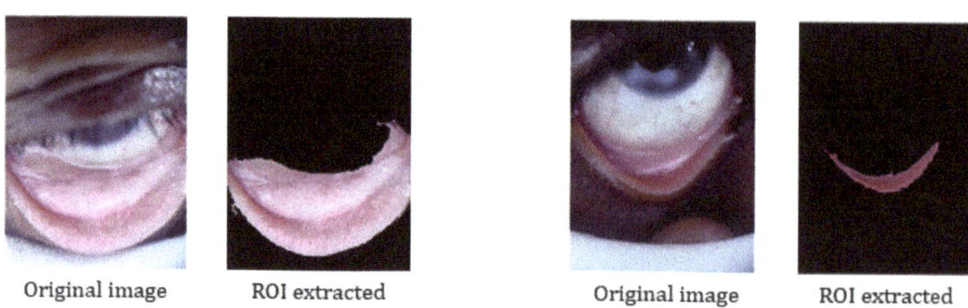

Figure 9. Examples of two images that would normally be discarded: the first because the eyelid overlaps the edge of the white spacer and is not perfectly in focus; additionally, the second one is not in focus and the finger appears to lower the eyelid. In both cases, automatic segmentation would still provide an acceptable result.

3. Results

The digital images of the patients' eyes have been captured by the device reported in Figure 2 and assembled on a Samsung S6 smartphone; 94 patients were involved, aged 19–75 (average 34), 46 female and 48 male, with Hb level concentrations in the range of 7.6–17.1 g/dL (average of 11.45 g/dL).

Each picture underwent a manual selection process, isolating and cropping regions of palpebral and forniceal conjunctiva, as shown in Figure 10. This step is needed to compare the manually segmented images considered as the ground truth with the automatic segmentation output from the proposed model. We evaluated both spatial and color properties of regions of interest by assessing the most suitable metrics based on this specific medical image segmentation problem [49]. F1 (FMS1), also known as the Sørensen–Dice coefficient, is the harmonic mean of precision and recall, defined as follows for binary segmentation applications:

$$F_1 = 2 \cdot (\frac{Precision \cdot Recall}{Precision + Recall}) = \frac{2 \cdot TP}{2 \cdot TP + FP + FN} \tag{8}$$

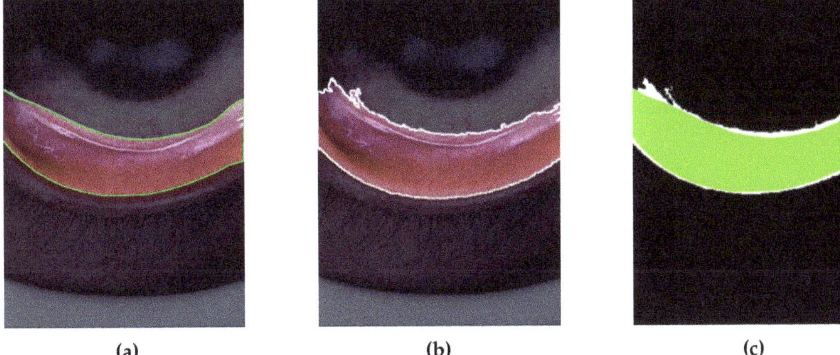

(a) (b) (c)

Figure 10. (**a**) Manually segmented conjunctiva used as ground truth. (**b**) Automatically segmented conjunctiva obtained by the proposed approach. (**c**) Visualization of the overlapping between green ground truth image and white automatically segmented image ($F1 = 0.904, accuracy = 96.41\%$).

The Dice coefficient being an overlapping measure ranging from 0 to 1, gives us a useful perspective about the quality of the segmentation. We are also interested in a calculation involving the number of pixels classified as non-relevant (false positive rate), which is not taken into account either by Dice coefficient or by Jaccard similarity. Accuracy metric is helpful in this case by outlining the rate of correctly classified pixels over the full image.

$$Accuracy = \frac{TP + TN}{TP + TN + FP + FN} \quad (9)$$

With the aim of assessing an average for the overlapping metrics, we computed a binary confusion matrix for each image. The values of this matrix refer to the number of pixels linked to set intersection or set difference between ground truth image and proposed segmentation, which are visually described in Figure 10c.

The averaged summation of each confusion matrix is summarized in Table 2. To give the reader the opportunity to observe the indicators for each sample included in the dataset, in Table A1 we have reported the values of the above metrics in a complete manner. Higher values of specificity for this segmentation task highlight the eligibility to disregard non conjunctival regions with proper confidence. On the other hand, sensitivity as well as F1 being overlapping measures, can reasonably fluctuate with higher variance, meaning in most cases that a finer meaningful subset of the conjunctival region has been selected.

Table 2. Metrics of averaged results of the comparison between manually and automatically segmented images of the conjunctiva.

	F1-Measure	Accuracy	Sensitivity (TPR)	Specificity (TNR)
Predicted ROIs	0.7363	93.79%	86.73%	94.63%

The optimal results indicated by the above metrics are sufficient to state the effectiveness of our segmentation algorithm. Since here we are dealing with a rigorous diagnostic procedure, if comparing the precision of the overlapping between proposed and ground truth ROIs is acceptable, we think that a further investigation of the color properties for left-out or added regions would be interesting.

CIELAB is one of the most useful amongst color spaces for erythema analysis and computer vision for diagnostics, composed by an approximately uniform three-dimensional space: L*, a*, b*. A widely used dimension from this space, a*, has a well-known correlation with hemoglobin values in this domain [36–38]. Our purpose is to examine the strength of linear correlation between mean values of a* extracted from digital images of conjunctivas and the relative Hb g/dL concentration from blood samples taken almost at the same time of picture capturing phase (Figure 11). Generalizing the idea of Pearson correlation coefficient (PCC) from two random variables to two standardized vectors, we can estimate the weight of their linear correlation ranging from −1 to 1 and defined by the following equation:

$$\rho(\boldsymbol{a}, \boldsymbol{b}) = \frac{1}{N-1} \sum_{i=1}^{N} \left(\frac{a_i - \mu_a}{\sigma_a}\right) \cdot \left(\frac{b_i - \mu_b}{\sigma_b}\right) \tag{10}$$

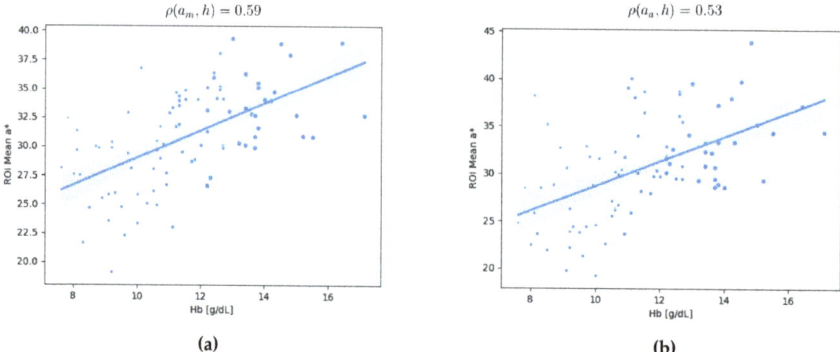

(a) **(b)**

Figure 11. (**a**) Linear regression and strength of correlation between a* from manual segmentation and Hb g/dL standardized vectors. (**b**) Linear regression and strength of correlation between a* from automatic segmentation and Hb g/dL standardized vectors.

We computed PCC between the mean a* values for both manually and automatically segmented images and Hb g/dL through the entire dataset of 94 samples, thereby obtaining respectively 0.59 and 0.53. The results reconfirm not only the moderate linear correlation between those values, but also a robust contiguity among human based manual segmentation and fully automated segmentation approach proposed.

4. Conclusions

We developed a fully automated segmentation procedure, based on graph partitioning, that exposes conjunctival regions while maximizing the correlation between color properties and hemoglobin concentration in the blood, according to the multi-layered anatomical structures of these tissues. The ROIs extracted by the model underwent an in-depth quantitative comparison with ground truth, using state of the art metrics for similarity and PCC between the a* component from CIELAB space and hemoglobin values. The results attest to the reliability and the capability of generalizing between patients belonging to heterogeneous classes, as the accuracy of the overlap between the manual and automatic ROIs selections, measured with classic metrics, is very good, and the correlation obtained between the level of Hb measured in vivo and that estimated through the color of the manual/automatic ROI are comparable. The proposed method paves the way for further studies involving deep learning techniques for both classifications of an estimated anemia risk category and regression to predict Hb real values. With this study we contribute to the broader diagnostic research field of image processing and analysis of the conjunctival pallor related to anemia diagnosis support. The advancement provided to this non-invasive image capturing procedure will lead to the possibility of embedding the model in a wearable device screening Hb risk category in real-time, without the need for physician support.

Author Contributions: The authors contributed equally to this work. All authors have read and agreed to the published version of the manuscript.

Funding: This research received no external funding.

Conflicts of Interest: The authors declare no conflict of interest.

Appendix A

Table A1. Results computed from the confusion matrices of the comparison between manually and automatically segmented images of the conjunctiva for the entire dataset of 94 samples.

Image ID	F1-Measure	Accuracy	Sensitivity (TPR)	Specificity (TNR)
164733	0.7547	0.9097	0.6060	1.0000
918410	0.7647	0.9287	0.6369	0.9567
094523	0.9123	0.9674	0.8696	0.9595
103722	0.6429	0.9687	0.6103	0.6792
190841	0.7011	0.8625	0.586	0.8724
154215	0.6494	0.9558	0.5505	0.7915
160737	0.7844	0.9327	0.6470	0.9957
155221	0.7179	0.9813	0.7044	0.7319
122613	0.7616	0.9176	0.8953	0.6627
132714	0.6641	0.8779	0.4971	1.0000
140525	0.7250	0.9316	0.9255	0.5959
154320	0.5296	0.8965	0.3602	1.0000
143315	0.7563	0.8955	0.6081	1.0000
145200	0.7834	0.9837	0.7677	0.7997
150240	0.6542	0.9170	0.4861	1.0000
155237	0.7672	0.9549	0.9374	0.6493
801000	0.7595	0.9460	0.9613	0.6277
121216	0.7848	0.9521	0.6534	0.9823
120556	0.6804	0.9080	0.5207	0.9815
134128	0.7827	0.9675	0.6715	0.938
150536	0.8229	0.9769	0.8237	0.8221
151234	0.7343	0.9285	0.6025	0.9400
155418	0.8351	0.9757	0.8186	0.8523
152136	0.7407	0.9264	0.8862	0.6362
152924	0.6875	0.9653	0.5282	0.9846
153536	0.6818	0.8958	0.5174	0.9995
154129	0.8665	0.9596	0.9719	0.7817
154759	0.8436	0.9559	0.7770	0.9226
155456	0.8463	0.9539	0.8111	0.8846
160045	0.6242	0.9333	0.4544	0.9965
123002	0.7943	0.9244	0.6703	0.9745
122915	0.7728	0.9664	0.7984	0.7488
232040	0.6222	0.9300	0.5065	0.8064
160522	0.8019	0.9790	0.8133	0.7909
121836	0.5998	0.8646	0.5157	0.7166
134745	0.7401	0.8944	0.7800	0.7040
211040	0.4881	0.9146	0.3258	0.9724
210631	0.9184	0.9838	0.9235	0.9134
223744	0.7676	0.9013	0.6468	0.9440

Table A1. *Cont.*

Image ID	F1-Measure	Accuracy	Sensitivity (TPR)	Specificity (TNR)
224452	0.655	0.8827	0.4872	0.9991
231923	0.7167	0.9513	0.5585	0.9999
232931	0.8046	0.9636	0.7029	0.9406
141804	0.7793	0.9310	0.7248	0.8428
152107	0.6693	0.9144	0.5063	0.9871
161452	0.7892	0.8955	0.7651	0.9673
154641	0.8193	0.9806	0.8627	0.7801
210419	0.8587	0.9675	0.8427	0.8753
221400	0.8056	0.9256	0.6767	0.9952
222325	0.8093	0.9298	0.6913	0.9758
140311	0.6237	0.9594	0.4608	0.9645
180148	0.8293	0.9154	0.7085	0.9998
183506	0.7559	0.9214	0.6226	0.9617
195511	0.7103	0.9149	0.5554	0.9849
201501	0.7197	0.9031	0.5662	0.9874
184029	0.7589	0.9715	0.6305	0.9531
184734	0.8508	0.9636	0.8814	0.8221
185602	0.8863	0.9722	0.8574	0.9172
190638	0.8229	0.9267	0.7120	0.9747
191233	0.8163	0.9388	0.8559	0.7801
191620	0.6685	0.8737	0.7922	0.5782
194457	0.7283	0.9508	0.5858	0.9624
114700	0.6357	0.9133	0.5007	0.8705
115146	0.6255	0.8800	0.5202	0.7842
115853	0.8018	0.9526	0.7490	0.8626
120426	0.6434	0.9588	0.5084	0.8762
202058	0.6903	0.8737	0.5271	1.0000
123714	0.7709	0.9415	0.8038	0.7406
133633	0.6015	0.9539	0.4604	0.8673
143301	0.8145	0.9803	0.7065	0.9614
144551	0.7174	0.9540	0.8865	0.6025
145301	0.6573	0.9124	0.4972	0.9693
150804	0.6424	0.9447	0.4849	0.9515
150539	0.8357	0.9547	0.7311	0.9750
151450	0.7388	0.9020	0.5886	0.9917
153146	0.7744	0.9295	0.6382	0.9844
162916	0.7940	0.9369	0.6713	0.9716
202947	0.9040	0.9641	0.8552	0.9587
180925	0.7136	0.9124	0.6152	0.8494
190130	0.8209	0.9776	0.7666	0.8834
190334	0.6594	0.9354	0.7855	0.5682
121621	0.8401	0.9549	0.7570	0.9436
154729	0.4816	0.9293	0.3244	0.9343
205012	0.8539	0.9651	0.9005	0.8120
205445	0.8337	0.9887	0.8632	0.8063
222551	0.7993	0.9394	0.7278	0.8863
223503	0.8563	0.9834	0.8353	0.8783

Table A1. *Cont.*

Image ID	F1-Measure	Accuracy	Sensitivity (TPR)	Specificity (TNR)
224240	0.7352	0.9379	0.6334	0.8760
205917	0.7118	0.9691	0.6264	0.8242
225922	0.7938	0.9492	0.8498	0.7447
231050	0.7480	0.9386	0.7003	0.8027
183626	0.5987	0.9463	0.4453	0.9133
161347	0.7855	0.9466	0.7371	0.8406
130148	0.6814	0.9690	0.5243	0.9728
130225	0.7896	0.9383	0.6632	0.9757

References

1. World Health Organization. *Worldwide Prevalence of Anaemia 1993–2005 : WHO Global Database on Anaemia*; de Benoist, B., McLean, E., Egli, I., Cogswell, M., Eds.; WHO: Geneva, Switzerland, 2008.
2. World Health Organization. *The World Health Report 2002*; World Health Organization: Geneva, Switzerland, 2002.
3. McLean, E.; Cogswell, M.; Egli, I.; Wojdyla, D.; Benoist, B. Worldwide prevalence of anaemia, WHO Vitamin and Mineral Nutrition Information System, 1993–2005. *Public Health Nutr.* **2008**, *12*, 444–54. [CrossRef] [PubMed]
4. Koury, M.J. Red blood cell production and kinetics. In *Rossi's Principles of Transfusion Medicine*; Wiley: Hoboken, NJ, USA, 2016, pp. 85–96. [CrossRef]
5. White, J.; Porwit, A.M. *Blood and Bone Marrow Pathology*; Elsevier: Amsterdam, The Netherlands, 2011.
6. World Health Organization; Centers for Disease Control and Prevention. *Assessing the Iron Status of Populations*; World Health Organization, Department of Nutrition for Health and Development: Geneva, Switzerland, 2005.
7. Marn, H.; Critchley, J.A. Accuracy of the WHO Haemoglobin Colour Scale for the diagnosis of anaemia in primary health care settings in low-income countries: A systematic review and meta-analysis. *Lancet Glob. Health* **2016**, *4*, e251–e265. [CrossRef]
8. Critchley, J.; Bates, I. Haemoglobin colour scale for anaemia diagnosis where there is no laboratory: A systematic review. *Int. J. Epidemiol.* **2005**, *34*, 1425–1434. [CrossRef] [PubMed]
9. Dimauro, G.; Girardi, F.; Gelardi, M.; Bevilacqua, V.; Caivano, D. Rhino-Cyt: A System for Supporting the Rhinologist in the Analysis of Nasal Cytology. *Intell. Comput. Theor. Appl. Lect. Notes Comput. Sci.* **2018**, 619–630. [CrossRef]
10. Dimauro, G.; Ciprandi, G.; Deperte, F.; Girardi, F.; Ladisa, E.; Latrofa, S.; Gelardi, M. Nasal cytology with deep learning techniques. *Int. J. Med Informatics* **2019**, *122*, 13–19. [CrossRef] [PubMed]
11. Triggiani, A.; Bevilacqua, V.; Brunetti, A.; Lizio, R.; Tattoli, G.; Cassano, F.; Soricelli, A.; Ferri, R.; Nobili, F.; Gesualdo, L.; et al. Classification of healthy subjects and Alzheimer's disease patients with dementia from cortical sources of resting state EEG rhythms: A study using artificial neural networks. *Front. Neurosci.* **2017**, *10*. [CrossRef]
12. Bevilacqua, V.; Pannarale, P.; Abbrescia, M.; Cava, C.; Paradiso, A.; Tommasi, S. Comparison of data-merging methods with SVM attribute selection and classification in breast cancer gene expression. *BMC Bioinform.* **2012**, *13*. [CrossRef]
13. Bevilacqua, V.; Cariello, L.; Columbo, D.; Daleno, D.; Fabiano, M.D.; Giannini, M.; Mastronardi, G.; Castellano, M. Retinal fundus biometric analysis for personal identifications. In Proceedings of the International Conference on Intelligent Computing, Shanghai, China, 5–18 September 2008; Springer: Berlin, Germany, 2008; pp. 1229–1237.
14. Bevilacqua, V.; D'Ambruoso, D.; Mandolino, G.; Suma, M. A new tool to support diagnosis of neurological disorders by means of facial expressions. In Proceedings of the IEEE International Symposium on Medical Measurements and Applications, Bari, Italy, 30–31 May 2011; pp. 544–549.

15. Dimauro, G.; Caivano, D.; Bevilacqua, V.; Girardi, F.; Napoletano, V. VoxTester, software for digital evaluation of speech changes in Parkinson disease. In Proceedings of the IEEE International Symposium on Medical Measurements and Applications (MeMeA), Benevento, Italy, 15–18 May 2016. [CrossRef]
16. Bevilacqua, V.; Brunetti, A.; Trotta, G.F.; Dimauro, G.; Elez, K.; Alberotanza, V.; Scardapane, A. A novel approach for Hepatocellular Carcinoma detection and classification based on triphasic CT Protocol. In Proceedings of the IEEE Congress on Evolutionary Computation (CEC), San Sebastian, Spain, 5–8 June 2017. [CrossRef]
17. Dimauro, G.; Nicola, V.D.; Bevilacqua, V.; Caivano, D.; Girardi, F. Assessment of Speech Intelligibility in Parkinson's Disease Using a Speech-To-Text System. *IEEE Access* **2017**, *5*, 22199–22208. [CrossRef]
18. Dimauro, G.; Caivano, D.; Girardi, F.; Ciccone, M.M. The patient centered Electronic Multimedia Health Fascicle-EMHF. In Proceedings of the IEEE Workshop on Biometric Measurements and Systems for Security and Medical Applications (BIOMS), Rome, Italy, 17 October 2014. [CrossRef]
19. Collings, S.; Thompson, O.; Hirst, E.; Goossens, L.; George, A.; Weinkove, R. Non-Invasive Detection of Anaemia Using Digital Photographs of the Conjunctiva. *PLoS ONE* **2016**, *11*, e0153286. [CrossRef]
20. Townsend, D.; D'Aiuto, F.; Deanfield, J. Super actinic 420 nm light-emitting diodes for estimating relative microvascular hemoglobin oxygen saturation. *J. Med Biol. Eng.* **2014**, *34*, 172–177. [CrossRef]
21. Zhao, Y.; Qiu, L.; Sun, Y.; Huang, C.; Li, T. Optimal hemoglobin extinction coefficient data set for near-infrared spectroscopy. *Biomed. Opt. Express* **2017**, *8*, 5151. [CrossRef] [PubMed]
22. Kim, O.; McMurdy, J.; Jay, G.; Lines, C.; Crawford, G.; Alber, M. Combined reflectance spectroscopy and stochastic modeling approach for noninvasive hemoglobin determination via palpebral conjunctiva. *Physiol. Rep.* **2014**, *2*, e00192. [CrossRef] [PubMed]
23. Sengupta, B. Biophysical Characterization of Genistein in Its Natural Carrier Human Hemoglobin Using Spectroscopic and Computational Approaches. *Food Nutr.* **2013**, *4*, 83–92.
24. Horecker, B. The absorption spectra of hemoglobin and its derivatives in the visible and near infra-red regions. *J. Biol. Chem.* **1943**, *148*, 173–183.
25. Sanchez-Carrillo, C. Bias due to conjunctiva hue and the clinical assessment of anemia. *J. Clin. Epidemiol.* **1989**, *42*, 751–754. [CrossRef]
26. Kent, A.; Elsing, S.; Hebert, R. Conjunctival vasculature in the assessment of anemia. *Ophthalmology* **2000**, *107*, 274–277. [CrossRef]
27. Kanchi, S.; Sabela, M.I.; Mdluli, P.S.; Inamuddin.; Bisetty, K. Smartphone based bioanalytical and diagnosis applications: A review. *Biosens. Bioelectron.* **2018**, *102*, 136–149. [CrossRef]
28. Escobedo, P.; Palma, A.J.; Erenas, M.M.; Olmos, A.M.; Carvajal, M.A.; Chavez, M.T.; Gonzalez, M.A.L.; Diaz-Mochon, J.J.; Pernagallo, S.; Capitan-Vallvey, L.F.; et al. Smartphone-Based Diagnosis of Parasitic Infections With Colorimetric Assays in Centrifuge Tubes. *IEEE Access* **2019**, *7*, 185677–185686. [CrossRef]
29. Ogirala, T.; Eapen, A.; Salvante, K.G.; Rapaport, T.; Nepomnaschy, P.A.; Parameswaran, A.M. Smartphone-based colorimetric ELISA implementation for determination of women's reproductive steroid hormone profiles. *Med Biol. Eng. Comput.* **2017**, *55*, 1735–1741. [CrossRef] [PubMed]
30. Wang, E.; Li, W.; Hawkins, D.; Gernsheimer, T.; Norby-Slycord, C.; Patel, S. HemaApp: Noninvasive Blood Screening of Hemoglobin Using Smartphone Cameras. *Getmobile: Mob. Comput. Commun.* **2017**, *21*, 26–30. [CrossRef]
31. Mannino, R.; Myers, D.; Tyburski, E.; Caruso, C.; Boudreaux, J.; Leong, T.; Clifford, G.; Lam, W. Smartphone app for non-invasive detection of anemia using only patient-sourced photos. *Nat. Commun.* **2018**, *9*. [CrossRef]
32. Sheth, T.; Choudhry, N.; Bowes, M.; Detsky, A. The Relation of Conjunctival Pallor to the Presence of Anemia. *J. Gen. Intern. Med.* **1997**, *12*, 102–106. [CrossRef] [PubMed]
33. Delgado-Rivera, G.; Roman-Gonzalez, A.; Alva-Mantari, A.; Saldivar-Espinoza, B.; Zimic, M.; Barrientos-Porras, F.; Salguedo-Bohorquez, M. Method for the Automatic Segmentation of the Palpebral Conjunctiva using Image Processing. In Proceedings of the IEEE International Conference on Automation/XXIII Congress of the Chilean Association of Automatic Control (ICA-ACCA), Concepcion, Chile, 17–19 Ocotber 2018; pp. 1–4.
34. Bevilacqua, V.; Dimauro, G.; Marino, F.; Brunetti, A.; Cassano, F.; Maio, A.D.; Nasca, E.; Trotta, G.F.; Girardi, F.; Ostuni, A.; et al. A novel approach to evaluate blood parameters using computer vision techniques. In Proceedings of the IEEE International Symposium on Medical Measurements and Applications (MeMeA), Benevento, Italy, 12–14 May 2016. [CrossRef]

35. Bauskar, S.; Jain, P.; Gyanchandani, M. A Noninvasive Computerized Technique to Detect Anemia Using Images of Eye Conjunctiva. *Pattern Recognit. Image Anal.* **2019**, *29*, 438–446. [CrossRef]
36. Dimauro, G.; Caivano, D.; Girardi, F. A new method and a non-invasive device to estimate anaemia based on digital images of the conjunctiva. *IEEE Access* **2018**, 1. [CrossRef]
37. Dimauro, G.; Guarini, A.; Caivano, D.; Girardi, F.; Pasciolla, C.; Iacobazzi, A. Detecting clinical signs of anaemia from digital images of the palpebral conjunctiva. *IEEE Access* **2019**, 1. [CrossRef]
38. Dimauro, G.; Baldari, L.; Caivano, D.; Colucci, G.; Girardi, F. Automatic Segmentation of Relevant Sections of the Conjunctiva for Non-Invasive Anemia Detection. In Proceedings of the 3rd International Conference on Smart and Sustainable Technologies (SpliTech), Split, Croatia, 26–29 June 2018; pp. 1–5.
39. Dhanachandra, N.; Manglem, K.; Chanu, Y.J. Image Segmentation Using K -means Clustering Algorithm and Subtractive Clustering Algorithm. *Procedia Comput. Sci.* **2015**, *54*, 764–771. [CrossRef]
40. Wu, M.N.; Lin, C.C.; Chang, C.C. Brain Tumor Detection Using Color-Based K-Means Clustering Segmentation. In Proceedings of the Third International Conference on Intelligent Information Hiding and Multimedia Signal Processing (IIH-MSP 2007), Kaohsiung, Taiwan, 26–28 November 2007. [CrossRef]
41. Chitade, A.; Katiyar, S. Color based image segmentation using K-means clustering. *Int. J. Eng. Sci. Technol.* **2010**, *2*, 5319–5325.
42. Shi, J.; Malik, J. Normalized Cuts and Image Segmentation. *IEEE Trans. Pattern Anal. Mach. Intell.* **2002**, *22*. [CrossRef]
43. Dimauro, G.; De Ruvo, S.; Di Terlizzi, F.; Ruggieri, A.; Volpe, V.; Colizzi, L.; Girardi, F. Estimate of Anemia with New Non-Invasive Systems—A Moment of Reflection. *Electronics* **2020**, *9*, 780. [CrossRef]
44. Tan, K.; Oakley, J. Enhancement Of Color Images In Poor Visibility Conditions. In Proceedings of the ICIP International Conference on Image Processing, Vancouver, BC, Canada, 10–13 September 2000. [CrossRef]
45. Arce, G.R.; Bacca, J.; Paredes, J.L. Nonlinear Filtering for Image Analysis and Enhancement. *Essent. Guide Image Process.* **2009**, 263–291. [CrossRef]
46. Graif, M.; Bydder, G.M.; Steiner, R.E.; Niendorf, P.; Thomas, D.; Young, I.R. Contrast-enhanced MR imaging of malignant brain tumors. *Am. J. Neuroradiol.* **1985**, *6*, 855–862. [PubMed]
47. Mammography, O.; Laine, A.; Fan, J.; Yang, W. Wavelets for Contrast Enhancement of Digital Mammography. *IEEE Eng. Med. Biol. Mag.* **1999**, *14*. [CrossRef]
48. Kaya, B.; Can, Y.B.; Timofte, R. Towards Spectral Estimation from a Single RGB Image in the Wild. *arXiv* **2018**, arXiv:cs.CV/1812.00805].
49. Taha, A.A.; Hanbury, A. Metrics for evaluating 3D medical image segmentation: Analysis, selection, and tool. *BMC Med. Imaging* **2015**, *15*. [CrossRef]

Article

An Efficient Hybrid Fuzzy-Clustering Driven 3D-Modeling of Magnetic Resonance Imagery for Enhanced Brain Tumor Diagnosis

Suresh Kanniappan [1], Duraimurugan Samiayya [1], Durai Raj Vincent P M [2], Kathiravan Srinivasan [2], Dushantha Nalin K. Jayakody [3,4,*], Daniel Gutiérrez Reina [5] and Atsushi Inoue [6]

1. Department of Information Technology, St. Joseph's College of Engineering, Chennai, Tamil Nadu 600119, India; sureshk@stjosephs.ac.in (S.K.); duraimurugans@stjosephs.ac.in (D.S.)
2. School of Information Technology and Engineering, Vellore Institute of Technology (VIT), Vellore, Tamil Nadu 632014, India; pmvincent@vit.ac.in (D.R.V.P.M.); kathiravan.srinivasan@vit.ac.in (K.S.)
3. School of Computer Science and Robotics, National Research Tomsk Polytechnic University, 634050 Tomsk, Russia
4. Centre for Telecommunication Research, Faculty of Engineering, Sri Lanka Technological Campus, Padukka 10500, Sri Lanka
5. Department of Electronic Engineering, University of Seville, 41092 Sevilla, Spain; dgutierrezreina@us.es
6. Information Systems and Business Analytics Department, Eastern Washington University, Spokane, WA 99202, USA; inoueatsushij@gmail.com

* Correspondence: nalin@tpu.ru

Received: 4 January 2020; Accepted: 5 February 2020; Published: 12 March 2020

Abstract: Brain tumor detection and its analysis are essential in medical diagnosis. The proposed work focuses on segmenting abnormality of axial brain MR DICOM slices, as this format holds the advantage of conserving extensive metadata. The axial slices presume the left and right part of the brain is symmetric by a Line of Symmetry (LOS). A semi-automated system is designed to mine normal and abnormal structures from each brain MR slice in a DICOM study. In this work, Fuzzy clustering (FC) is applied to the DICOM slices to extract various clusters for different k. Then, the best-segmented image that has high inter-class rigidity is obtained using the silhouette fitness function. The clustered boundaries of the tissue classes further enhanced by morphological operations. The FC technique is hybridized with the standard image post-processing techniques such as marker controlled watershed segmentation (MCW), region growing (RG), and distance regularized level sets (DRLS). This procedure is implemented on renowned BRATS challenge dataset of different modalities and a clinical dataset containing axial T2 weighted MR images of a patient. The sequential analysis of the slices is performed using the metadata information present in the DICOM header. The validation of the segmentation procedures against the ground truth images authorizes that the segmented objects of DRLS through FC enhanced brain images attain maximum scores of Jaccard and Dice similarity coefficients. The average Jaccard and dice scores for segmenting tumor part for ten patient studies of the BRATS dataset are 0.79 and 0.88, also for the clinical study 0.78 and 0.86, respectively. Finally, 3D visualization and tumor volume estimation are done using accessible DICOM information.

Keywords: MR brain segmentation; fuzzy clustering; object extraction; silhouette analysis; DICOM processing; 3D modeling

---

1. Introduction

Brain tumor detection is crucial in medical diagnosis as it provides adequate information about anomalies present in the tissues. This information is necessary to understand the prognosis of the disease and also for treatment planning [1]. Magnetic Resonance Imaging (MRI) procedures help to sense the irregularities of human bodies in three dimensions, non-invasively. In particular, various segmentation techniques are applied to MR brain images by radiographers to identify the extent of abnormality present [2,3]. Recently, many Computer-Aided Detection (CAD) methods are employed for brain tumor detection [4–6]. Subsequently, radiologists anticipate that usage of CAD schemes over brain MR images can advance the diagnostic capabilities with their collaborative effects [7,8].

The Digital Imaging and Communications in Medicine (DICOM) standard image format delivers increased diagnostic relevance. DICOM-compliant MR imaging devices adhere to a specific protocol for archiving and communication of digital medical images. DICOM (.dcm) files afford metadata information such as patient study, equipment settings, and image characteristics-modality, size, bit depth, and dimensions. The DICOM header object is organized as a standard series of tags. These tags are categorized as groups such as image pixel, the image plane, MR/CT Image, and patient information [9,10].

The size of this header differs depending on data elements in each group. For, eg: the image plane module contains various vital parameters, which include image position, slice location, and pixel spacing. From these parameters, the spatial relationship between the slices is computed. DICOM facilitates to create private tags that define data elements accessed within the application created. Various imaging modalities stores digital images in DICOM format, which provides better volume of metadata compared to other formats. DICOM provides harmonization through which the patient under study is wholly analyzed and it also compatible with many commercial toolkits.

The patient dataset is inherently acquired by DICOM-compliant devices. Many methods had been proposed by researchers to segment desired features from the digital images. Intensity-based segmentation methods rely on fixing thresholds and are easier to implement [11]. However, due to high-intensity variations of MR images, the methods yield poor performance and lacks in piecewise continuity. Clustering techniques are standard iterative algorithms that is based on the minimization of an objective function. It considers the pixel intensity values for precisely classifying the image pixels. The extraction of cells or tissues based on morphology, clustering algorithms are used extensively. Many algorithms existing in the literature have the objective to yield better segmentation. With the K-means clustering algorithm [12], a large set of structures is distributed into disjoint and homogeneous clusters. Dhanachandra has attempted image segmentation using a hybrid combination of K-means clustering and the Subtractive Clustering Algorithm [13]. Abdel-Maksoud attempted a combined approach of K-means and Fuzzy C-means clustering technique for brain tumor detection [14]. Kim proposed quantization of full/partial (thickness) tear of rotator cuff tendon using Fuzzy C-Means based classification [15]. Dehariya proposed the segmentation of images using Fuzzy K-means clustering [16]. Gasch implemented Fuzzy k-means clustering as an analytical tool for mining biological perceptions from yeast gene-expression data [17]. Even clustering techniques perform faster computation, a wrong choice of k may produce inaccurate results.

Markov random fields (MRFs) benefit more straightforward implementation by encoding spatial data which expresses a set of parameters for specifying tumor voxels [18,19]. This method is very robust for MR images and their performance entirely depends on spatial constraints and hence not suitable for heterogeneous tissue classes. Statistical pattern recognition based methods also known as atlas-based segmentation methods, are effective only for bi-level segmentation. These approaches require healthier brain atlas that is modified significantly to accommodate the tumor part which may lead to poor results. Hybrid methods utilize the advantage of many models which is used in numerous applications by integrating different models within a system to enhance segmentation accuracy. Fuzzy clustering exhibits excellent performance on images containing homogeneous and heterogeneous tissue classes [20]. However, fuzzy clustering produces better results by choosing the proper selection of the number of clusters 'k'.

In the literature, to assess the number of clusters, a metric-based method called silhouette score is used. It evaluates the number of clusters based on their proximity. The silhouette score is interpreted as excellent, moderate, weak and bad splits based on cluster selection. Lleti had attempted to optimize the silhouettes using a genetic algorithm in choosing variables for the K-means cluster examination [21]. Muca determined the optimal number of clusters based on the silhouette index for the K-means algorithm [22]. Robust segmentation based on the finest silhouette scores is performed on a set of DICOM slice sequences that assists in the segregation of abnormal portions from the brain tissue.

Numerous approaches have been proposed for the detection of various objects of interest after segmentation is performed. Zeng proposed K-means with a hybrid active contour model to generate an initial segmentation for segmenting thick-vessels in liver images [23]. Koulountzios developed a simple pipeline for segmenting the whole thoracic aorta into contours such as arch, descending, and ascending aorta from MR DICOM files containing thoracic region [24]. Nekooeimehr proposed a method for tracking and segmenting organ contours using k-means clustering with prior information [25]. Wang has attempted contour refinement using an active contour model to segregate candidate cavernoma sections from brain MR slices [26]. An improved performance utilizing local and global image information for contour detection into a hierarchical region tree [27]. Essadike suggested Van der Lugt correlator-based initial contour to assist an active contour model in extracting tumor boundaries [28].

Morphology is a broad set of non-linear operations that process images that rely on shape and texture classification [29,30]. Ali attempted the K-Means Clustering technique for accounting pixel intensities and locations [31]. The author had applied to dilate and erode morphological operations to abstract the tumor part from the brain tumors, which also aided to eliminate small isolated points. Deng employed morphological operators to enhance the extracted ulcer area from ocular staining images [32].

A comparative investigation between the mined region of interest (ROI) and master segmented (Ground Truth) images is carried out with the well-known image similarity measures [33,34]. The Jaccard and Dice coefficients are calculated to validate the segmentation performed on each slice against their corresponding ground truth object. Modeling 3D view of a patient study requires resampling and image interpolation methods [35] to align the abnormal intensities in the spatial domain geometrically.

The key contributions of this work are summarized as follows:

- This research study uses the advantage of fuzzy clustering (for image enhancement) hybridized with Distance Regularized Level Set technique to effectively mine the region of interest form the brain slices.
- In this work, for each brain slice we have utilized the attributes of DICOM standards such as Image position patient, Pixel spacing and Image orientation patient, which is essential for generating the 3D model of brain structures and volumetric analysis.
- For image enhancement in identifying the objects of interest, fuzzy clustering is employed through proper selection of the number of clusters 'k' validated using the silhouette metric. The appropriate k is chosen based on the silhouette metric among the number of clusters (k) ranging from 2 to 9.
- The proposed work is initially tested on the brain MR series of BRATS dataset for anomaly extraction; its segmentation quality is assessed with image quality, similarity and statistical measures. The average dice scores over ten patient studies for tumor segmentation has given promising results. Further, the procedure is also tested on the clinical MR brain series and validated against expert ground truth.

In this work, the proposed tool is implemented using python open-source language. The proposed methodology is described in Section 2. The obtained results and their relevant findings are demonstrated in Section 3. The conclusions and future scope are discussed in Section 4. A brief video describing the proposed method, its key contributions' and results, is provided in the Supplementary Materials.

2. Materials and Methods

Two different brain slice datasets were used in this work. Firstly, the real-world clinical dataset, which comprises 22 brain slices (axial T2 MR DICOM slices), obtained from the Proscans Diagnostics Centre (Chennai, India). Secondly, the benchmark BRATS dataset was used for evaluating the performance of the proposed model. Further, in this work, the BRATS dataset comprised of ten patients and around 200 brain slices were acquired from each patient. This section specifies that the proposed approach was deployed to segment and analyze the axial MR DICOM slices. Initially, the DICOM slices are subjected to pre-processing. The segmentation of preprocessed DICOM slices is subjected to fuzzy clustering for image enhancement. In order to select the best clusters, the silhouette metric is employed. The enhancement of extracted structures is carried out using morphological operations. Finally the ROI is extracted using image post-processing procedures such as MCW, RG, and DRLS, and the extracted tumor is validated using similarity measures. The complete architecture is shown in Figure 1. Also volumetric quantification of tumor and 3D visualization is generated from the slices involved in the real-time clinical study. The decision making capability of the proposed approach is tested and validated using 2D slices of the considered image dataset.

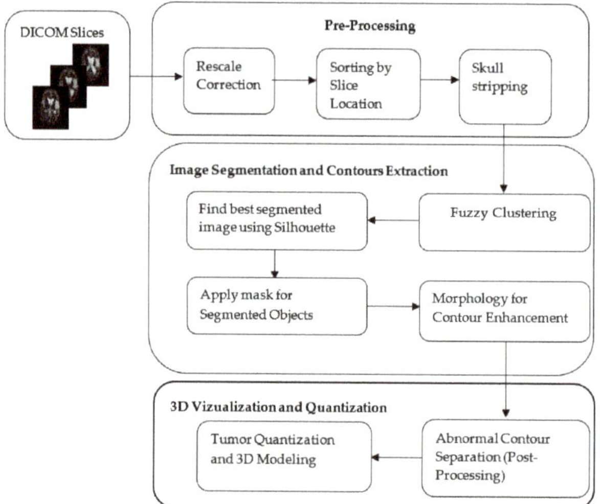

Figure 1. The Architecture of the Proposed Hybrid Model.

2.1. Pre-Processing

The considered slices are presented to the system as (.dcm) files. Rescale correction is performed on all the slices which provide a 512*512 pixel array for each image. The DICOM tags Rescale Intercept (RI) and Rescale Slope (RS) postulates the linear transformation of pixels to their memory representation. The Rescale Correction [36] of the slices is given by:

$$RC = I{*}RS + RI \tag{1}$$

where RC is the rescaled units, I is the intensity value. In MR DICOM metadata, the attributes rescale slope and rescale intercept are not available as tags. For the computational purpose, the tag values are engaged as one and 1024, respectively. The available slice location header attribute in MR allows brain slices to be added for processing sequentially.

Abnormality identification in the brain requires removal of the skull or non-brain tissues such as dura, arachnoid, pia mater for effective extraction of ROI. The skull portions possess a low solidity area. Solidity is the proportion of the contour area to its convex hull area. Regions having the least solidity are removed, leaving only the region occupied by the actual brain tissue. The slices which have high solidity objects were retained after rescaling correction, thereby eliminating the non-brain matter.

2.2. Fuzzy Clustering Based Object Extraction from DICOM Slices

The fuzzy clustering (FC) technique is performed to extract the best segmentation in a feature space containing varying cluster intensities and shapes. Fuzzy clustering classifies a set of data points $p_1, p_2, p_3, \ldots, p_m$ of a DICOM slice into k ($\leq m$) clusters, which minimizes the total distortion. Each data point in the feature space has a degree of membership (a_{ij}) to its own cluster. The points closest to cluster centroids has a higher degree compared to the points in the cluster edge. For a data point i assigned to cluster j, gives a_{ij} coefficient value for being in the jth cluster. The sum of a_{ij} coefficient is always 1. The fuzzy assisted clustering algorithm based on minimization of the following objective function (F_w) concerning A (fuzzy k partition) and B (k set of clusters) and it is given by:

$$F_w(A,B) = \sum_{j=1}^{m}\sum_{i=1}^{k}\left(a_{ij}\right)^w d^2\left(P_j, B_i\right);\ k \leq m \tag{2}$$

where, $w(>1)$, is the weighting exponent acts as a control parameter for the fuzziness in a_{ij}, P_j is the jth point in the feature vector of N-dimensional space, B_i being the centroid for cluster i, a_{ij} is the degree of membership of the pixel P_j in cluster i, $d^2\left(P_j, B_i\right)$ is the distance measure between P_j and B_i, m and k represent the number of data points and the number of clusters, respectively.

The degree of membership of all feature vectors is associated with the inverse of the distance to the cluster center:

$$a_{ij} = \frac{\left[\frac{1}{d^2(P_j,B_i)}\right]^{1/w-1}}{\sum_{i=1}^{k}\left[\frac{1}{d^2(P_j,B_i)}\right]^{1/w-1}} \tag{3}$$

The Euclidean distance measure is used to compute the degree of membership (a_{ij}) is given as:

$$d^2\left(P_j, B_i\right) = \left(P_j - B_i\right)^T I\left(P_j - B_i\right) \tag{4}$$

where I denotes the identity matrix. The new centroid positions are computed based on the mean of all the points, weighted by its corresponding degree of membership (a_{ij}) to the cluster:

$$\widehat{B_i} = \frac{\sum_{j=1}^{n}\left(a_{ij}\right)^w P_j}{\sum_{j=1}^{n}\left(a_{ij}\right)^w} \tag{5}$$

Based on new centroid positions the updated degree of membership $\left(\widehat{a_{ij}}\right)$, is computed according to a_{ij} shown in Equation (3). This process is repeated until the sum of distances of each point in the slice to the centroid of the cluster is minimum, i.e., a termination criterion '$\in$' is reached, which ensures maximum accuracy. Other stopping criteria include no further improvement in the variance over some iterations. The structures of the brain are segmented from the set of DICOM slices for a set of 'k' values ranging from 2 to 9.

2.3. Selection of the Best (k) Using Silhouette Index

Fuzzy clustering renders the clustered image for the preferred number of clusters (k). However, optimal 'k' should be chosen in order to place cluster labels within the centroid. In literature, a well-balanced coefficient named silhouette score, presented by Rousseeuw [37], has shown higher performance in finding optimal clusters. The silhouette score pertains to the deviation between the within-class tightness and separation. Specifically, the silhouette value for a pixel in the slice pixel array is given by,

$$sil(j) = \frac{b(j) - a(j)}{max(a(j), b(j))} \tag{6}$$

where, $a(j) = 1/|V_i| - 1\sum_{y \in V_i} d(x, y)$, be the mean distance of pixel point 'x' with other pixels (y) within the cluster V_i and $b(j) = min\ \{1/V_i \sum_{y \in V_i} d(x, y)\}$, be the average dissimilarity of a point 'x' to about any cluster V_i of which chosen point 'y' is not associated with it.

The maximum value of $s(j)$ reflects the optimal number of clusters. Correspondingly, the minimum of $b(j)$ is taken for computing $s(j)$. If $b(j)$ is larger, then the point is very far from its next neighboring cluster. The squared Euclidean distance provides the distance metric $d(x, y)$ between clusters for computing the silhouettes.

K-means clustering with silhouette analysis is executed to find out the optimal 'k' ranging from 2 to 9.

Silhouette always lies between −1 to 1, and it is defined as:

$$sil(j) = \begin{cases} 1 - a(j)/b(j),\ if\ a(j) < b(j) \\ 0\ ,\ if\ a(j) = b(j) \\ b(j)/a(j) - 1,\ if\ a(j) > b(j) \end{cases} \tag{7}$$

If the silhouette values are approaching either +1 or −1, the pixel points are well clustered or misclassified, respectively. If zero, the points could be assigned to another cluster also.

Further, to validate the segmented slices, the entire pixel array of each slice is considered, and the average silhouette width is computed. The average silhouette width for every slice is calculated from mean of all the distinct cluster silhouettes is given by:

$$S_{avg}(s_i) = \frac{\sum_i^n sil(j)}{k} \tag{8}$$

where, n denotes the number of clusters segmented. The S_{avg} is used to find the best k for a slice s_i. The silhouette coefficient ($k_{best,i}$) is defined as the maximum average silhouette width which is given by,

$$k_{best,i} = max\{S_{avg}(s_i)\} \tag{9}$$

The algorithmic steps of incorporating fuzzy clustering and silhouette metric to the set of DICOM slices are illustrated in Algorithm 1.

Algorithm 1 Silhouette-enabled Fuzzy Clustering

```
Let S = {s_1, s_2, s_3, ..., s_m}                  (Set of Dicom Slices)
P = {p_1, p_2, p_3, ..., p_m}                      (Set of data points to be clustered)
K_r, r ∈ cluster_range [2:10]
k_best = Best K value of the clustered image
B = {b_1, b_2, b_3, ..., b_k}            (Set of cluster Centroids)
for each s_i ∈ S
    for each k in K_r
        for each p_i ∈ P
            Compute fuzzy Clustering by iteratively updating the degree of membership (â_ij) and cluster
            centroids B̂_i
        end
for every k in S_i
Compute averagesilhouettewidth from individual cluster silhouettes (finding best k from r.)
end
Compute k_best,i = max(averagesilhouettewidth)  (Calculate k_best for Slice i)
End
```

2.4. Morphological Operations for Objects Enhancement

Image masking is used to specify the foreground, background, or probable background/foreground. Contour masking separates the objects from the original images, and it is essential for further analysis. It is eliminating the outliers such as air, from the actual brain slices. The fuzzy clustering process discovers the best-segmented clusters. These clusters form a binary mask that overlaid on the actual slices to acquire the respective contour intensities. Mutual information (I) is computed between the contour mask with the corresponding slice for ensuring similarity [38,39]. The weighted contribution (Wi) of the contour mask (CMi) to the original slice (Si) is calculated as:

$$W_i = \frac{1}{CM_i} e^{-\frac{I_i - I_{min}}{I_{max} - I_{min}}} \tag{10}$$

where I_i represents the mutual information between CM_i and S_i. I_{max}, I_{min} are the maximum and minimum mutual information for the overall CM_i. After the extraction of structures from the fuzzy clustering process, the obtained binary mask of the chosen slices may be distorted due to noise and texture. Mathematical morphology, a kind of contrast enhancement technique, assists selective enhancement of the small diagnostic contour features that are overlaid on a composite background. Hence, the binary mask representing the extracted structures is further practiced with non-linear operations such as morphological erode and morphological dilate for removing the inadequacies in order to retain the form and structure of the extracted objects. Erosion is a reverse process to dilation-erosion strips pixel layer over the edges, contradictorily dilation augments pixel layer over the edges.

Dilation adds pixels to the contour boundaries in the slices. The number of supplementary pixels integrated into the mask image is subject to the shape and size of the structuring element. Dilation process is done by:

$$CM \oplus SE = \{z|(\widehat{SE})_z \cap CM \neq \varphi\} \tag{11}$$

where CM is the set of pixels representing the binary mask, SE be the structuring element initially reflected as $\widehat{SE}$ then the reflected element is translated by z. This process enlarges the binary mask in all directions not to miss any pixels, particularly at contour edges. Similarly, erosion is performed by:

$$CM \ominus SE = \left\{\left\{z\middle|(SE)_z \subseteq CM\right\}\right\} \tag{12}$$

Stating that z confined in CM shifts the SE. Erosion removes pixels, thus sharpening the object boundary. The number of pixels stripped is subject to the size of the SE. Erosion strips the connected normal and abnormal contours, which aid in the effective extraction of ROI in the post-processing stages.

2.5. Tumor Quantization and Validation

The extracted objects possess high solidity ventricles and the tumor region. In order to extract the ROI from the brain structures, image post-processing approaches are utilized to mine the ROI from the extracted objects. After determining the abnormal regions, the size of the tumor is quantified based on its area and perimeter.

The validation metrics are used to evaluate the spatial intersection of ground truth (GT) of the clinical slices with the extracted ROI [40]. The performance of the segmentation procedure is validated using similarity measures such as Dice, Jaccard, false positive (FPR), and false negative (FNR) rates.

These measures are mathematically conveyed as:

$$Jaccard\left(I_{gt}, I_{ROI}\right) = (I_{gt} \cap I_{ROI})/(I_{gt} \cup I_{ROI}) \tag{13}$$

$$Dice\left(I_{gt}, I_{ROI}\right) = 2(I_{gt} \cap I_{ROI})/(I_{gt} \cup I_{ROI}) \tag{14}$$

$$FPR\left(I_{gt}, I_{ROI}\right) = (I_{gt}/I_{ROI})/(I_{gt} \cup I_{ROI}) \tag{15}$$

$$FNR(I_{gt}, I_{ROI}) = (I_{ROI}/I_{gt})/(I_{gt} \cup I_{ROI}) \quad (16)$$

where, I_{gt} expresses to the ground truth (GT) and I_{ROI} points for the segmented image with the proposed strategy. Other related works implemented on brain MRI can be found in [41–49].

2.6. Volume Assessment and 3D Modeling

The clinical slices considered in this work have the cubical stack format [SC × W × H], SC signifies the number of slices (22), W, and H indicates the width and height of a slice [512 × 512] in pixels, respectively. The slices are processed in DICOM format, which holds adequate slice information. In DICOM metadata, it is identified that 'slice thickness (ST)' is 5 mm, 'pixel spacing (PS_x/PS_y)' is (0.4492 mm/0.4492 mm) and 'spacing between slices (SS)' is 6.5 mm. The original stack [22 × 512 × 512] migrated to a resampled stack [110 × 230 × 230] based on the spacing information presented in the DICOM header. Resampling is achieved by cubic spline interpolation function. The tumor volume is estimated as:

$$Volume = Tumor\ Voxels * Voxel\ size \quad (17)$$

where tumor voxels are the number of voxels that contributes to tumor and voxel size is measured based on pixel spacing and slice thickness.

Visualization of tumor voxels in the complete study is achieved by merging the tumor containing slices to form a voxel mesh in all three anatomical planes. Further, this exemplifies an interconnected set of triangular faces of tumor voxels [50].

3. Results and Discussions

This section illustrates the results achieved with the proposed procedure. Figure 2 demonstrates a brief overview of the proposed procedure for brain MR examination. Initially, the DICOM slices of the patient study are pre-processed with rescaling correction. Moreover, this creates the intensity of the similar tissues in the study to confirm across the image slices. Then the non-brain tissues are stripped from the brain matter, leaving the brain pixels, which contain the brain tissues. Figure 3 shows the representative image of slice 12 before extraction (a) of brain tissue after removing (b) the non-brain portions. This procedure avoids non-brain tissues to add unnecessary information, thereby enhancing the efficacy of extracting the ROI. The combination of Fuzzy clustering with validated silhouette index (as the criterion for choosing best k) discovers the precise number of clusters from the slices. Figure 4 shows the extracted objects of Slice-14 and its corresponding mask image for '*k*' clusters (*k* ranges from 2 to 9).

Table 1 shows the silhouette scores S_{avg} for the considered slices for the chosen k values. A more substantial silhouette value gives a high split over the data points. For slice10 in Table 1, the optimal k is elected as two, since $S_{avg}(s_i)$ holds the maximum value 0.45571. Similarly, the optimal $k_{best,i}$ is selected from the range of slices (Slice11-Slice14) based on the average silhouette width.

Table 1. Silhouette scores for Slices (10–14).

Slices	k = 2	k = 3	k = 4	k = 5	k = 6	k = 7	k = 8	k = 9
Slice 10	**0.45571**	0.37847	0.3974	0.40063	0.40432	0.40568	0.411	0.35408
Slice 11	0.42842	0.41451	**0.44141**	0.43435	0.43878	0.4262	0.43644	0.43349
Slice 12	0.44109	0.45273	0.4498	0.46001	**0.47158**	0.4616	0.46032	0.46397
Slice 13	0.47107	0.50479	**0.51933**	0.50445	0.51069	0.50843	0.50041	0.43237
Slice 14	0.48988	0.53767	0.54796	0.55355	**0.58523**	0.53146	0.52066	0.52847

The maximum silhouette score ($k_{best,i}$) obtained for the cluster range are highlighted in **bold**.

In slice14 the maximum $S_{avg}(s_{14}) = 0.58523$ when *k* is six, but a minimum value resulted when *k* is two, ($S_{avg}(s_{14}) = 0.48988$). Further, this shows the loss of tumor information in Figure 4 when *k* is two, and an optimal segmentation is obtained when *k* is six.

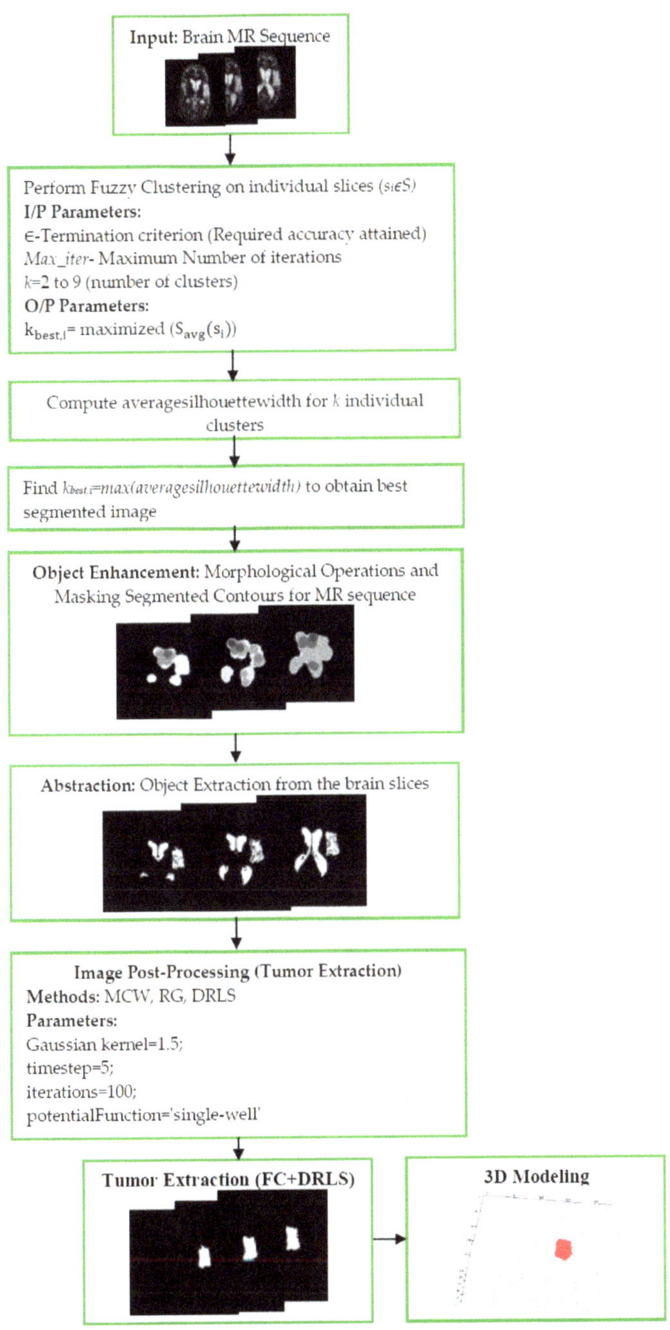

Figure 2. Flowchart of the proposed brain MRI examination of a patient study.

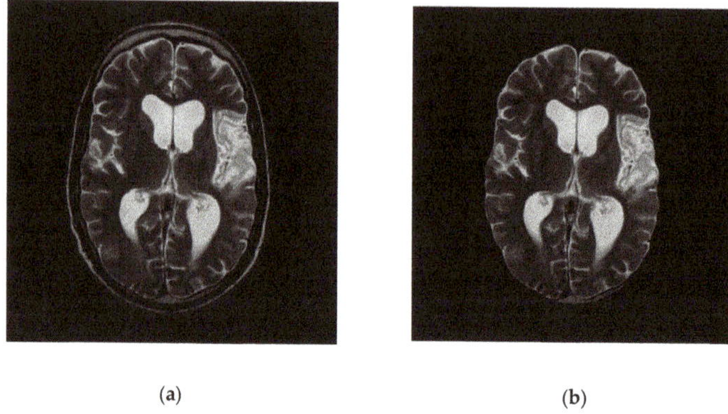

Figure 3. Slice (12) Before (**a**) and after (**b**) skull stripping.

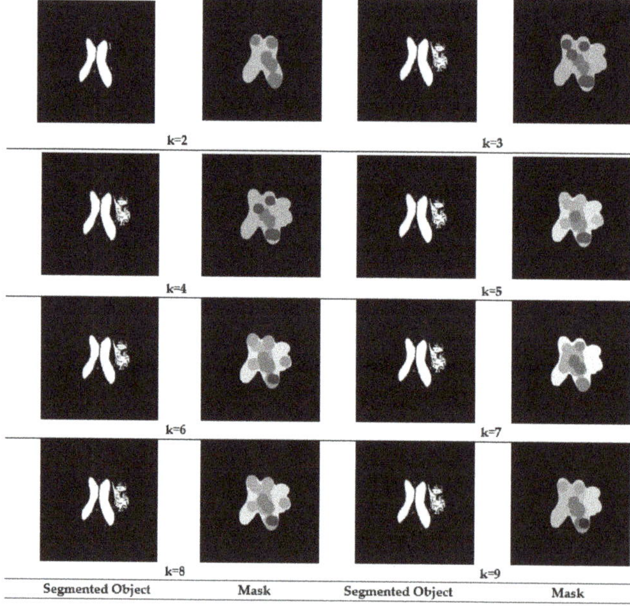

Figure 4. Segmented results and Masked Objects for Slice 14.

Figure 5a shows the graph with silhouette scores versus the number of clusters '*k*' for a representative sample image 'slice14.dcm'. The average silhouette width for the slices, which are more than 0.48, confirms a good split for all *k* clusters. As shown in the graph, a reasonable peak is obtained when $k = 6$ and also possess a maximum value ($S_{avg}(s_{12}) = 0.58523$).

Figure 5b depicts the silhouette plot for slice14. The selection of an optimal *k* provides better separation in feature space with more similar thickness and sizes. This increase in $S_{avg}(s_{14})$ is due to the distinct separation from their neighboring clusters except for the background. The well-diversified information is obtained from the slices since the proper selection silhouette index was made. From the achieved outcomes, it is clear that silhouette analysis plays a significant role in the identification of best-clustered objects. After FC, the pixels representing abnormalities are spatially identified and segmented. The mask image is produced from the objects extracted from each slice, in order to distinguish from outliers/ background. The mask image holds the pixel intensity values of abnormal and normal pixel intensities. The mask is applied to the corresponding slice for the extraction of clustered intensities from the actual slices. The objects are further enhanced by morphological processes by performing morphological open, dilate, erode and close operations in a sequential order to obtain a smoother object boundary without speckles. The experimental results of Figure 3 confirms a superior image enhancement step; as a result, it shows the best separation of objects obtained from composite backgrounds.

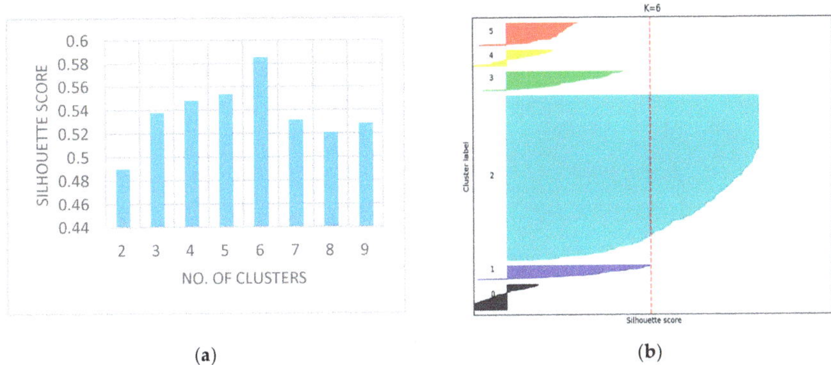

(**a**) (**b**)

Figure 5. (**a**) No. of Clusters vs. Silhouette Scores; (**b**) Silhouette plot for Slice-14.

Each slice in the MR sequence is distinctly examined using the eminent state-of-art segmentation methods such as watershed [51], Chan-Vese [52], and fuzzy clustering (FC). The image quality measures obtained from these methods are portrayed in Table 2. From the segmented results of FC, it is perceived that FC based segmentation offers distinct separation of objects and aids a better confidant for the image post-processing stages.

Table 2. Segmented results for state-of-art segmentation methods.

	Slice 10	Slice 11	Slice 12	Slice 13	Slice 14
Watershed					
Chan-Vese					
Fuzzy Clustering					

Further, for measuring the supremacy of the segmentation approaches, the well-known image quality measures [53] such as peak signal to noise ratio (PSNR), normalized cross-correlation (NCC), normalized absolute error (NAE) and structural similarity index (SSIM) are calculated and are presented in Table 3. From the table values, it is perceived that the FC method discovers the vital prominent structures and thus preserves the segmentation quality. Also, the average image quality measures attained through FC is superior compared to the other state-of-art approaches reflected in this work.

Table 3. Image quality measures for segmentation methods.

Segmentation	Slice	PSNR	NCC	NAE	SSIM
Watershed	Slice 10	21.3173	0.4923	0.6852	0.7955
	Slice 11	19.8605	0.3846	0.7571	0.7095
	Slice 12	22.6953	0.4742	0.7021	0.6992
	Slice 13	20.8065	0.4192	0.7121	0.7148
	Slice 14	21.7631	0.4031	0.7231	0.7219
	Average	21.28854	0.43468	0.71592	0.72818
Chan-vese	Slice 10	24.0187	0.5102	0.6712	0.7083
	Slice 11	23.0823	0.5802	0.6328	0.7153
	Slice 12	22.0176	0.4979	0.6693	0.6983
	Slice 13	25.2131	0.5374	0.6501	0.7213
	Slice 14	23.0129	0.5278	0.6712	0.6859
	Average	23.46892	0.5307	0.65892	0.70582
Fuzzy Clustering	Slice 10	20.9234	0.4865	0.7091	0.6995
	Slice 11	28.6764	0.7681	0.4065	0.8204
	Slice 12	30.5289	0.7548	0.3773	0.8143
	Slice 13	32.3411	0.7917	0.3961	0.8968
	Slice 14	31.5401	0.7842	0.3843	0.8412
	Average	28.80198	0.71706	0.45466	0.81444

The proposed approach had been tested for its performance by validating it against the grand challenge benchmark image dataset called the BRATS (size: 236 × 216 pixels). In this dataset, ten patient studies of T2 and T1C modalities had been taken up for analysis, which contains axial brain MR image series. The sample image series of patient studies and their corresponding GT that are obtained are shown in Figure 6. The BRATs dataset that had been considered in the proposed research has a number of advantages, few of those are–The desirable amount of 2D slices of a patient study can be easily extracted from its skull stripped 3D brain MRI, modalities like Flair, T1, T1C, and T2 are easily supported, contains ground truth images for all modalities offered by an expert member. Due to these reasons that most researchers had adopted the BRATS images for testing their disease examination tool.

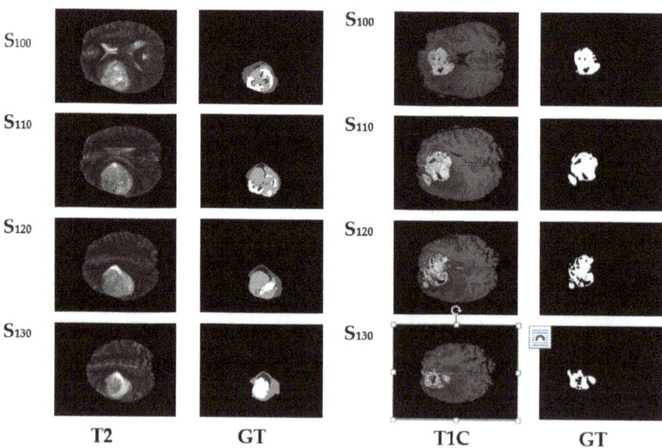

Figure 6. BRATS Dataset image series with Expert's Ground Truth.

For a comparative analysis in the post-processing stage, the ROI mining technique, marker controlled watershed segmentation (MCW), Seed region growing (RG) and distance regularized level sets (DRLS) are adopted and implemented to extract the anomalous section from the segmented objects (shown in Figure 4). MCW is a well-known segmentation technique associated with marker controlled morphological function and Sobel's edge detection. This procedure detects ROI with the assistance of priori provided whole alike image intensities [54].

In contrast to MCW, the level sets the implicit active contour models, uses gradient information of the image, and thus naturally handles topological deviations by merging or splitting the contours [55]. The parameters for DRLS is assigned as follows; number of iterations = 100, scale parameter = 1.5, potential function = single-well and timestamp = 5.

RG is an operator instigated semi-automated method, extensively used to extract the desired (abnormal) structures from medical images [56]. RG requires a seed point (pixel position) to be initialized somewhere within a contour or ROI. From the seed point, the RG procedure will start mining the similar intensities of possible connected neighboring pixels accessible in the ROI.

The images chosen are initially subjected to FC for objects enhancement before applying the post-processing. The image quality gets enhanced through the validated FC technique and provides an ideal platform for the post-processing that is performed using MCW, RG, and DRLS. Figure 7 depicts the brain abnormality segmentation results extracted from the 2D slices of T2 modality images through FC assisted MCW, RG, and DRLS techniques. Correspondingly, results of segmented ROI from 2D slices of T1C modality are shown in Figure 8.

The segmentation methods (MCW, RG, and DRLS) that had been implemented were assessed for their performance by carrying out a comparative analysis that was executed between the ROI and GT. The extracted ROI and GT were initially compared on T2 modality images, followed by T1C modality images. The results obtained from these comparisons were recorded in Tables 4 and 5. The recordings were made based on image similarity measures like Jaccard, Dice, FPR, and FNR. Figure 9 shows the comparative analysis of assessed similarity measures of T2 and T1C weighted images. The corresponding average scores of Tables 4 and 5 are depicted in the graph.

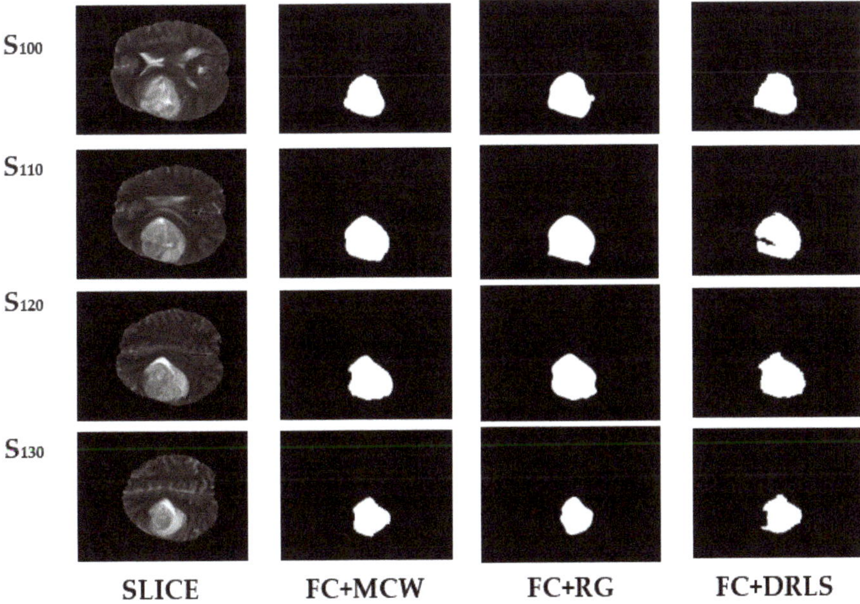

Figure 7. Segmentation results of BRATS T2 series.

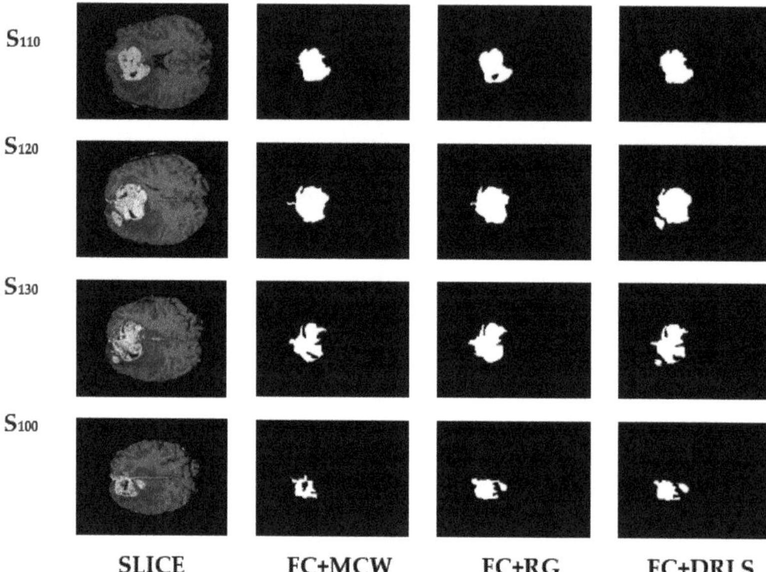

Figure 8. Segmentation results of BRATS T1C series.

Table 4. Image similarity measures for BRATS T2 MRI series.

	Slice	Jaccard	Dice	FPR	FNR
FC+MCW	S_{100}	0.8625	0.9201	0.00617	0.0487
	S_{110}	0.7623	0.8587	0.00126	0.1327
	S_{120}	0.7162	0.828	0.00427	0.1117
	S_{130}	0.7428	0.7048	0.002044	0.2123
FC+RG	S_{100}	0.8914	0.9421	0.00796	0.0742
	S_{110}	0.7785	0.8691	0.00127	0.1566
	S_{120}	0.7189	0.8299	0.00144	0.1951
	S_{130}	0.7478	0.7868	0.00923	0.3995
FC+DRLS	S_{100}	0.8958	0.9334	0.00288	0.083
	S_{110}	0.7939	0.9091	0.00119	0.179
	S_{120}	0.7427	0.8525	0.00176	0.1925
	S_{130}	0.7592	0.7951	0.00159	0.2636

Table 5. Image similarity measures for BRATS T1C MRI series.

	Slice	Jaccard	Dice	FPR	FNR
FC+MCW	S_{100}	0.6645	0.8284	0.0047	0.1904
	S_{110}	0.5154	0.9067	0.0064	0.1868
	S_{120}	0.6923	0.8643	0.0069	0.1628
	S_{130}	0.7187	0.8322	0.0061	0.1954
FC+RG	S_{100}	0.7071	0.9123	0.0051	0.1962
	S_{110}	0.8293	0.8045	0.0058	0.1895
	S_{120}	0.761	0.8999	0.0052	0.1552
	S_{130}	0.7127	0.9034	0.0043	0.2001
FC+DRLS	S_{100}	0.7628	0.8655	0.0054	0.2467
	S_{110}	0.8376	0.9116	0.0057	0.0994
	S_{120}	0.8187	0.9003	0.0059	0.1567
	S_{130}	0.7357	0.8477	0.0054	0.1886

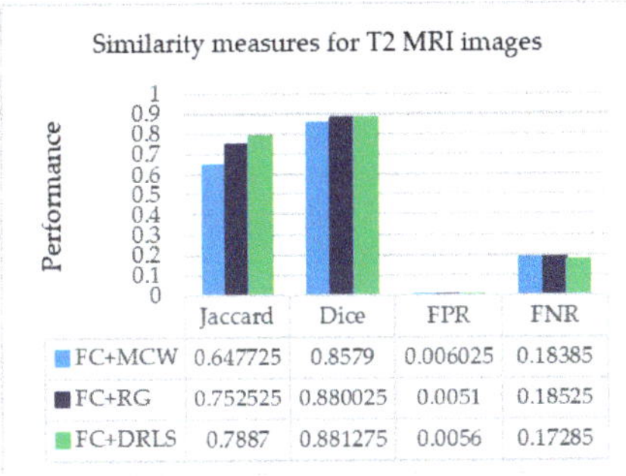

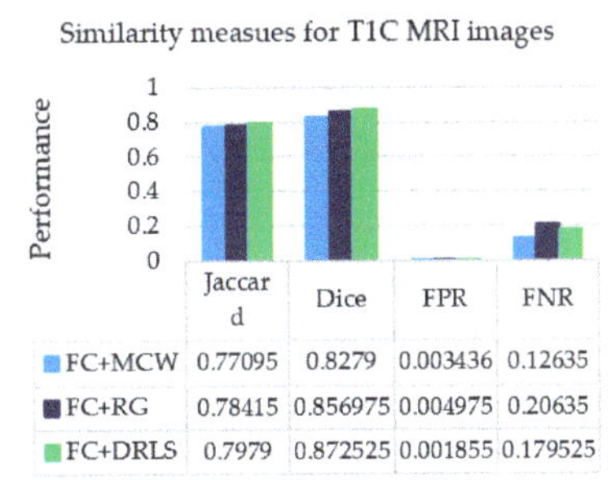

Figure 9. Average picture similarity measures of a patient study against expert's ground truth.

From the recorded values for the considered slices presented in Tables 4 and 5, it could be inferred that the outcome produced through FC based DRLS technique is far more superior to FC+MCW and FC+RG techniques. Also, the metrics Jaccard and Dice are computed for ten patient studies of BRATS individually. Figures 10 and 11 depicts the average scores for ten patient studies of the BRATS dataset, and 'Av.' represents the overall average score of all the patient studies from T2 and T1C weighted images, respectively.

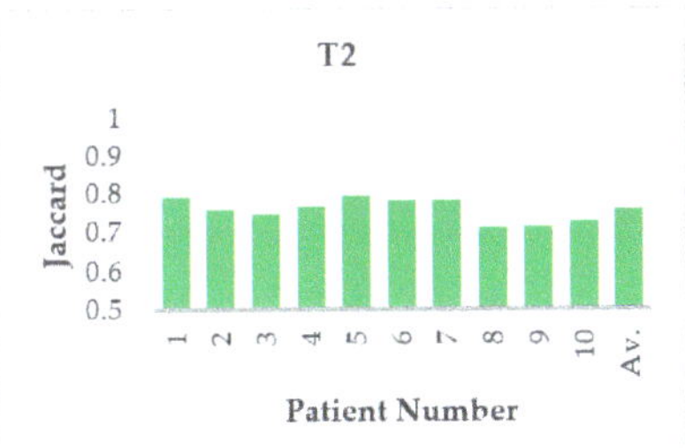

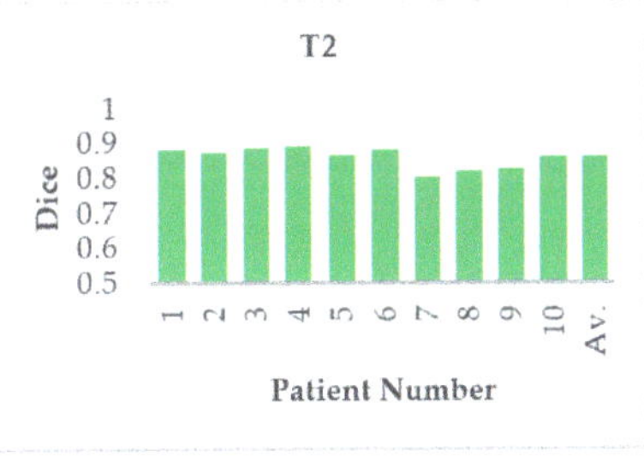

Figure 10. Average Jaccard and Dice score of individual patient studies for T2 modality. 'Av.' specifies the average score of all the patient studies.

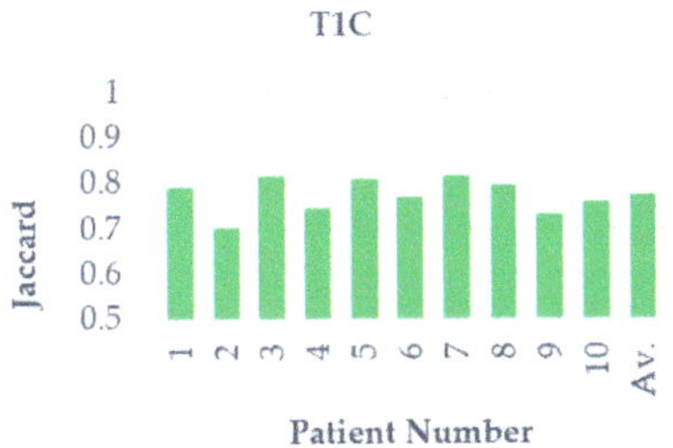

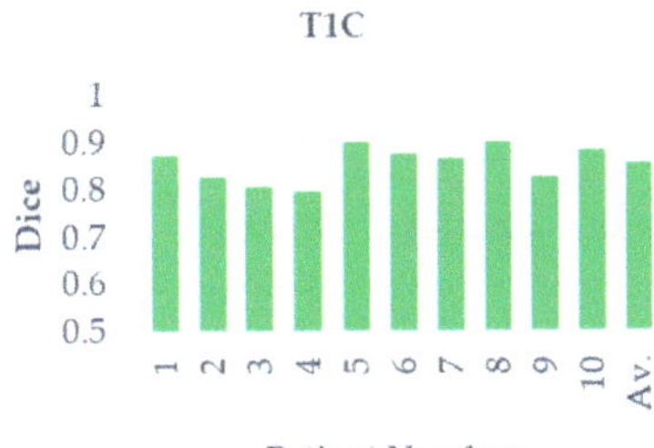

Figure 11. Average Jaccard and Dice score of individual patient studies for T1C modality. 'Av.' Specifies the average score of all the patient studies.

Furthermore, the suggested procedure is attempted on clinical MRI brain study of a patient [57]. The patient study considered contains axial T2 MR DICOM slices. The DICOM slices between the ranges Slice-10 to Slice-14 are considered in this approach for tumor analysis, as they hold enough tumor information. Slice15 and above are excluded from the examination as it does not contain any tumor region. Initially, the considered series are enhanced using FC, then at the post-processing stage, the mining techniques MCW, RG, and DRLS are adopted. Figure 12a,b embody the slice number with an optimal k and the original (actual) middle slices. Figure 12c represents the ground truth provided by an expert member. Figure 12d–f signify the ROI extracted from the validated fuzzy clustering-assisted MCW, RG, and DRLS procedures.

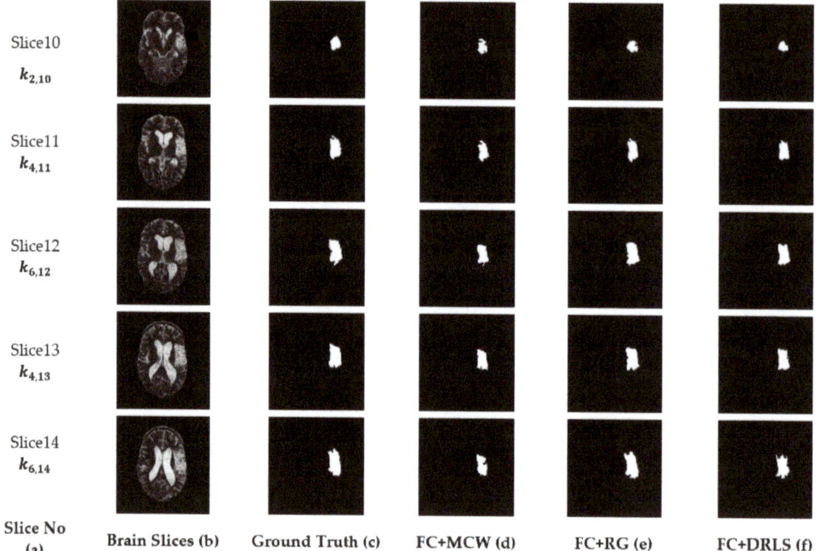

Figure 12. Segmented results of the clinical dataset (middle slices only).

The validation of the FC-aided mining procedures against GT images is performed using well-known image similarity measures such as Dice, Jaccard, false-negative, and false-positive rates. These parameters stay as an aid to assess the efficacy of the segmentation procedure.

From Table 6, it is observed that maximum similarity is attained between ROI and GT by the suggested FC+DRLS procedure shown on the Jaccard and Dice metric scores. FPR values indicate that typical pixels of the brain are misclassified as tumor pixels. Similarly FNR values depict pixels which contribute as tumor are misclassified as normal pixels of the brain.

Table 6. Image similarity measures for clinical T2 MRI series.

	Slice	Jaccard	Dice	FPR	FNR
FC+MCW	S_{10}	0.6932	0.7832	0.1027	0.0612
	S_{11}	0.8223	0.8828	0.1075	0.0311
	S_{12}	0.8185	0.8569	0.1579	0.0424
	S_{13}	0.8137	0.8761	0.1683	0.0422
	S_{14}	0.7314	0.8168	0.1544	0.1201
FC+RG	S_{10}	0.7209	0.7767	0.1123	0.0723
	S_{11}	0.8149	0.8934	0.1099	0.0793
	S_{12}	0.7953	0.8491	0.1184	0.0683
	S_{13}	0.8054	0.7962	0.1163	0.0923
	S_{14}	0.7821	0.8021	0.1201	0.0876
FC+DRLS	S_{10}	0.6874	0.7949	0.1336	0.0642
	S_{11}	0.8104	0.8972	0.1253	0.032
	S_{12}	0.8256	0.8625	0.1368	0.0296
	S_{13}	0.8179	0.8879	0.1374	0.038
	S_{14}	0.7555	0.8596	0.1164	0.0958

The minimum values of FPR and FNR guarantee the efficiency of the FC+DRLS segmentation method against MCW and RG in the set of DICOM slices. Figure 13 shows the comparative analysis of assessed similarity measures using the average scores of the slices depicted in Table 6. Therefore it is evident that in the proposed approach, the FC+DRLS based segmentation technique produces superior results for the clinical study as well.

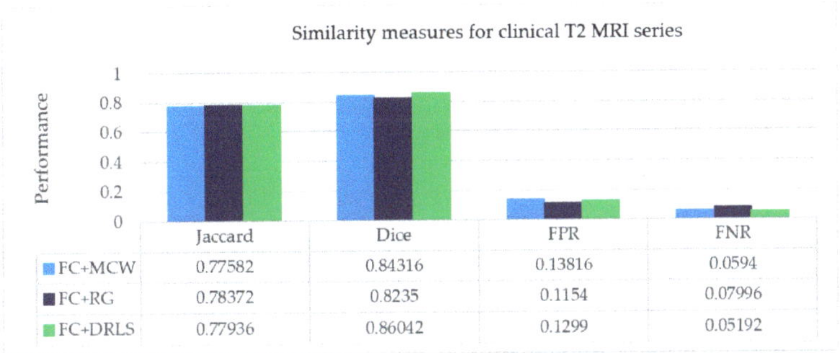

Figure 13. Average picture similarity measures of clinical study with expert's ground Truth.

Further, the ROI (tumor) is extracted to acquire the geometrical properties such as area and perimeter. The captured tumor information is clearly visible from slice10 to slice14. In Table 7, the parameters, area, and perimeter are calculated based on the tumor information extracted from the ROI. The parameters gradually increase up to slice13, which holds the maximum tumor part and then decreases. The overall study of a patient is determined from the entire DICOM slices.

Table 7. Quantization features Area and Perimeter for best 'K'.

Slices	FC Enhancement	ROI by DRLS	Area	Perimeter
Slice 10 (k = 2)			2268	192.5097
Slice 11 (k = 4)			4842	324.60
Slice 12 (k = 6)			6090	366.74
Slice 13 (k = 4)			6091	375.90
Slice 14 (k = 6)			6008	320.1680
(a)	**(b)**	**(c)**	**(d)**	**(e)**

The slices are reconstructed to a cubical stack based on in-stack position attribute of DICOM. The extracted objects of the clinical study undergo volumetric estimates and 3D reconstruction. Table 8 shows the volume of tumor calculated for DICOM and resampled stack.

Table 8. Volume calculation of DICOM and Resampled stack.

Stack	No. of Voxels	Voxel Size	Volume (mm^3)
DICOM	21623	1.0089032	21,815.5
Resampled	26911	0.9998	26,905.6

Tumor volume is calculated for the DICOM grid using the physical spacing metadata available in the DICOM header. The inter-slice resolution of the considered patient study is coarse, as the slice thickness is 5 mm, which is considerably higher than the in-plane pixel size, i.e., 0.4492 mm. This anisotropic characteristic results in appalling issues for modeling 3D and image analysis. Thus resampling is often considered as a vital step to transform DICOM stack to an isotropic stack. In the resampled stack, the accuracy of integrating the contours of individual slices not degraded; also, it interpolates the z dimension with lower resolution and in-plane dimensions with higher resolution.

Figure 14 shows the 3D models of DICOM and resampled grid for the patient study. The three anatomical planes are used, and their dimensions are set in mm. For the DICOM grid in Figure 14a, it is observed that the dimensions [X × Y × Z] are set as [512 × 512 × 22]. Similarly, the resampled grid in Figure 14b the dimensions are viewed as [230 × 230 × 110]. The model permits to visualize the object interactively in all the three directions specified.

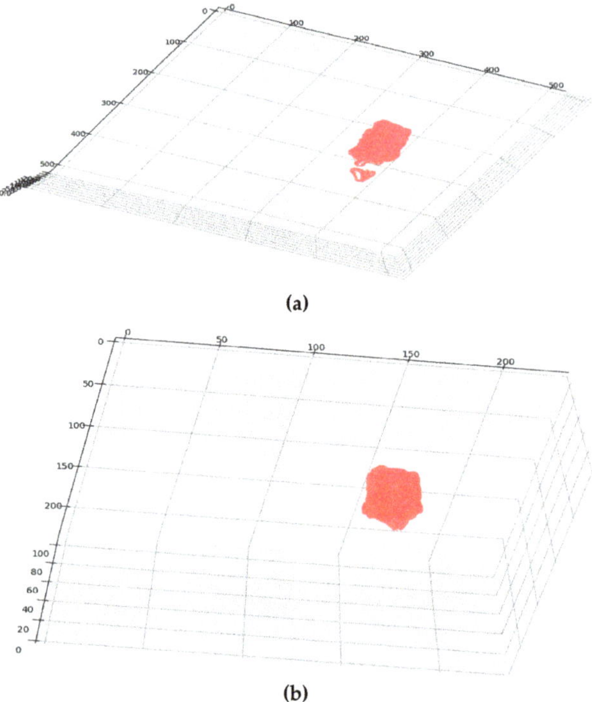

Figure 14. (**a**) DICOM Grid; (**b**) Resampled Grid.

From Figure 14a,b, it is clear that the resampled grid offers a smoother surface than the DICOM grid. Resampled grid offers a smoother iso-surface and better 3D visualization.

The proposed work had been validated against various modalities of the BRATS dataset and clinical slices. The performance progression is carried out at each stage of the suggested segmentation procedure. The fuzzy clustering technique had given prominent results in the enhancement phase that aids in effective extraction in post-processing stages. The proposed work had considered two to nine classes (k) that were applied on each slice of the patient under study for ascertaining the most prominent k that could yield the best segmentation. The silhouette score is taken as validation metric results in the optimal enhancement of slices since it considers the $k_{best,i}$ measure for making up the number of required classes. This validated clustering process helps in minimizing the loss of tumor intensities over the patient study. Also, a comparative segmentation analysis had been carried out against Chan-Vese and watershed algorithms for ensuring the segmentation quality of FC. To overcome the computational complexity, the proposed work had considered $k = 9$ as the upper limit for the number of clusters. For DRLS post-processing, imparting single well potential function and Gaussian kernel value as 1.5 had yielded better extraction of tumor part than RG and MCW techniques. In the future, the proposed procedure can be pondered on brain slices containing diffused boundaries and other image modalities in addition to magnetic resonance angiograms (MRA).

4. Conclusions

In this work, a hybrid procedure is implemented, which uses fuzzy clustering with silhouette analysis followed by MCW, RG, and DRLS procedures. Moreover, this proposed method applied to the entire slices of abnormal patient studies obtained from the BRATS challenge and the Proscans Diagnostics Centre. This investigation delivered better segmentation of the regions where the concentration of tumor was high. The best-segmented objects are obtained using clustering techniques which are further evaluated by silhouette metrics. The tumor objects from the enhanced slices are segmented based on MCW/RG/DRLS techniques. The quantification results of the mined anomalies ensure the progression of counterpart tumors at different treatment stages. The clinical significance of the proposed hybrid approach gives a better prognosis identification against the ground truth. The use of python open source technologies in implementing the work can visualize, analyze and interact with the slice data claim to be cost-effective. Hence, the proposed framework on MR DICOM slices requires less user intervention in extracting tumor heterogeneity from typical brain structures. Quantification and 3D modeling procedure help in finding a spatial identity and tumor concentration. By knowing the size, shape and spatial location of the tumor, the process of treating the tumor might be improved. The future work could include the implementation of advanced artificial intelligence methodologies for early, efficient, and real-time diagnosis of malignant brain tumors [58–65].

Supplementary Materials: The video abstract can be found at the following link: https://drive.google.com/file/d/1ddr0DxPNP1cX7aMC-dvSx12sQJ4fuH59/view?ts=5e452180. Video: An Efficient Hybrid Fuzzy-Clustering Driven 3D-Modeling.

Author Contributions: Conceptualization, S.K., D.S.; methodology, S.K., D.S., D.R.V.P.M.; software, S.K.; validation, K.S. D.N.K.J., D.G.R.; formal analysis, S.K., D.S., D.R.V.P.M.; investigation, S.K., D.S., D.R.V.P.M.; resources, K.S.; data curation, S.K., K.S.; writing—original draft preparation, S.K.; writing—review and editing, D.S., D.R.V.P.M., K.S., D.N.K.J., D.G.R., A.I.; visualization, D.N.K.J., D.G.R.; supervision, K.S., D.N.K.J., and project administration, K.S., D.N.K.J.; funding acquisition, D.N.K.J. All authors have read and agreed to the published version of the manuscript.

Funding: This work was funded, in part, by the Scheme for Promotion of Academic and Research Collaboration (SPARC), Ministry of Human Resource Development, India under the SPARC/2018-2019/P145/SL, in part, by the framework of Competitiveness Enhancement Program of the National Research Tomsk Polytechnic University, Russia in part, and, in part, by the International cooperation project of Sri Lanka Technological Campus, Sri Lanka and Tomsk Polytechnic University, Russia, No. RRSG/19/5008.

Acknowledgments: The authors would like to acknowledge the support granted by Proscans Diagnostics Centre, the leading and reputed Pathology Lab network in Chennai, Tamilnadu, India, for providing real clinical images of the brain MRI.

Conflicts of Interest: The authors declare that they have no conflict of interest.

Ethical Approval: This article follows the ethical standards of 1964 Helsinki declaration with its future amendments.

References

1. Karaa, W.B. *Biomedical Image Analysis and Mining Techniques for Improved Health Outcomes*; IGI Global: Hershey, USA, 2015. [CrossRef]
2. El-Dahshan, E.S.; Mohsen, H.M.; Revett, K.; Salem, A.B.M. Computer-aideddiagnosis of human brain tumor through MRI: A survey and a new algorithm. *Expert Syst. Appl.* **2014**, *41*, 5526–5545. [CrossRef]
3. Chyzhyk, D.; Savio, A.; Graña, M. Evolutionary ELM wrapper feature selection for Alzheimer's disease CAD on anatomical brain MRI. *Neurocomputing* **2014**, *128*, 73–80. [CrossRef]
4. Virmani, J.; Dey, N.; Kumar, V. PCA-PNN, and PCA-SVM based CAD systems for breast density classification. In *Applications of Intelligent Optimization in Biology and Medicine*; Springer: Berlin/Heidelberg, Germany, 2016; pp. 159–180.
5. Bahadure, N.B.; Ray, A.K.; Thethi, H.P. Image analysis for MRI based brain tumor detection and feature extraction using biologically inspired BWT and SVM. *Int. J. Biomed. Imag.* **2017**, *2017*. [CrossRef]
6. Fujita, H.; Uchiyama, Y.; Nakagawa, T.; Fukuokab, D.; Hatanakac, Y.; Hara, T.; Lee, G.N.; Hayashi, Y.; Ikedoa, Y.; Gaoa, X.; et al. Computer-aided diagnosis: The emerging of three CAD systems induced by Japanese health care needs. *Comput. Methods Progr. Biomed.* **2008**, *92*, 238–248. [CrossRef]
7. Marshkole, N.; Singh, B.K.; Thoke, A.S. Texture and shape-based classification of brain tumors using linear vector quantization. *Int. J. Comput. Appl.* **2011**, *30*, 21–23.
8. Arimura, H.; Tokunaga, C.; Yamashita, Y.; Kuwazuru, J. Magnetic resonance image analysis for brain CAD systems with machine learning. In *Machine Learning in Computer-Aided Diagnosis: Medical Imaging Intelligence and Analysis*; IGI Global: Hershey, USA, 2012; pp. 258–296. [CrossRef]
9. Bidgood Jr, W.D.; Horii, S.C.; Prior, F.W.; Van Syckle, D.E. Understanding and using DICOM, the data interchange standard for biomedical imaging. *J. Am. Med. Inf. Assoc.* **1997**, *4*, 199–212. [CrossRef]
10. Herrmann, M.D.; Clunie, D.A.; Fedorov, A.; Doyle, S.W.; Pieper, S.; Klepeis, V.; Le, L.P.; Mutter, G.L.; Milstone, D.S.; Schultz, T.J.; et al. Implementing the DICOM standard for digital pathology. *J. Pathol. Inform.* **2018**, *9*, 1–18. [CrossRef]
11. Suresh, K.; Sakthi, U. Robust multi-thresholding in noisy grayscale images using Otsu's function and harmony search optimization algorithm. In *Advances in Electronics, Communication and Computing*; Springer: Singapore, 2018. [CrossRef]
12. Hartigan, J.A.; Wong, M.A. Algorithm AS 136: A k-means clustering algorithm. *J. R. Stat. Soc. Ser. C* **1979**, *28*, 100–108. [CrossRef]
13. Dhanachandra, N.; Manglem, K.; Chanu, Y.J. Image segmentation using K-means clustering algorithm and subtractive clustering algorithm. *Proc. Comput. Sci.* **2015**, *54*, 764–771. [CrossRef]
14. Abdel-Maksoud, E.; Elmogy, M.; Al-Awadi, R. Brain tumor segmentation based on a hybrid clustering technique. *Egypt. Inf. J.* **2015**, *16*, 71–81. [CrossRef]
15. Kim, K.B.; Song, Y.S.; Park, H.J.; Song, D.H.; Choi, B.K. A fuzzy C-means quantization based automatic extraction of rotator cuff tendon tears from ultrasound images. *J. Intell. Fuzzy Syst.* **2018**, *35*, 149–158. [CrossRef]
16. Dehariya, V.K.; Shrivastava, S.K.; Jain, R.C. Clustering of image data set using k-means and fuzzy k-means algorithms. In Proceedings of the 2010 International Conference on Computational Intelligence and Communication Networks, Bhopal, India, 26–28 November 2010; pp. 386–391. [CrossRef]
17. Gasch, A.P.; Eisen, M.B. Exploring the conditional coregulation of yeast gene expression through fuzzy k-means clustering. *Genome Biol.* **2002**, *3*, research0059-1. [CrossRef] [PubMed]
18. Rajinikanth, V.; Raja, N.S.M.; Kamalanand, K. Firefly algorithm assisted segmentation of tumor from brain MRI using Tsallis function and Markov random field. *J. Control Eng. Appl. Inf.* **2017**, *19*, 97–106.
19. Raja, N.S.M.; Lakshmi, P.R.V.; Gunasekaran, K.P. Firefly algorithm-assisted segmentation of brain regions using tsallis entropy and Markov random field. In *Innovations in Electronics and Communication Engineering*; Springer: Singapore, 2018; pp. 229–237.
20. Gath, I.; Geva, A.B. Unsupervised optimal fuzzy clustering. *IEEE Trans. Pattern Anal. Mach. Intell.* **1989**, *515*, 87–100. [CrossRef]
21. Lletı, R.; Ortiz, M.C.; Sarabia, L.A.; Sánchez, M.S. Selecting variables for k-means cluster analysis by using a genetic algorithm that optimises the silhouettes. *Anal. Chim. Acta* **2004**, *515*, 87–100. [CrossRef]

22. Muca, M.; Kutrolli, G.; Kutrolli, M. A proposed algorithm for determining the optimal number of clusters. *Eur. Sci. J.* **2015**, *11*, 36.
23. Zeng, Y.Z.; Liao, S.H.; Tang, P.; Zhao, Y.Q.; Liao, M.; Chen, Y.; Liang, Y.X. Automatic liver vessel segmentation using 3D region growing and hybrid active contour model. *Comput. Biol. Med.* **2018**, *97*, 63–73. [CrossRef]
24. Koulountzios, P.I.; Zervakis, M.E.; Karakitsios, P.L.; Stavroulakis, G.E. A semi-automatic algorithm for reconstruction and NURBS surface generation of thoracic aorta. In Proceedings of the 2017 IEEE International Conference on Imaging Systems and Techniques (IST), Beijing, China, 20 October 2017; pp. 1–6. [CrossRef]
25. Nekooeimehr, I.; Lai-Yuen, S.; Bao, P.; Weitzenfeld, A.; Hart, S. Automated contour tracking and trajectory classification of pelvic organs on dynamic MRI. *J. Med. Image.* **2018**, *5*, 014008. [CrossRef]
26. Wang, H.; Ahmed, S.N.; Mandal, M. Computer-aided diagnosis of cavernous malformations in brain MR images. *Comput. Med. Image Gr.* **2018**, *66*, 115–123. [CrossRef]
27. Arbelaez, P.; Maire, M.; Fowlkes, C.; Malik, J. Contour detection and hierarchical image segmentation. *IEEE Trans. Pattern Anal. Mach. Intell.* **2010**, *33*, 898–916. [CrossRef]
28. Essadike, A.; Ouabida, E.; Bouzid, A. Brain tumor segmentation with Vander Lugtcorrelator based active contour. *Comput. Meth. Prog. Biomed.* **2018**, *160*, 103–117. [CrossRef] [PubMed]
29. Hemalatha, S.; Anouncia, S.M. A computational model for texture analysis in images with fractional differential filter for texture detection. *Int. J. Ambient Comput. Intell.* **2016**, *7*, 93–113. [CrossRef]
30. Hu, S.; Wang, X.; Zhu, M.; Hao, G.; Yao, C.; Hu, C.-H. Differentiation of High-grade Gliomas from Brain Metastases Using Tissue Similarity Maps (TSMs) Based Relative Cerebral Blood Volume Values. *Curr. Med. Image Rev.* **2018**, *14*, 594–598. [CrossRef]
31. Ali, S.M.; Abood, L.K.; Abdoon, R.S. Brain tumor extraction in MRI images using clustering and morphological operations techniques. *Int. J. Geogr. Inf. Syst. Appl. Remote Sens.* **2013**, *4*, 12–25.
32. Deng, L.; Huang, H.; Yuan, J.; Tang, X. Automatic segmentation of corneal ulcer area based on ocular staining images. In *Proceedings of the Medical Imaging 2018: Biomedical Applications in Molecular, Structural, and Functional Imaging*; International Society for Optics and Photonics: Houston, TX, USA, 2018. [CrossRef]
33. Dey, N.; Rajinikanth, V.; Ashour, A.S.; Tavares, J.M.R. Social group optimization supported segmentation and evaluation of skin melanoma images. Symmetry. *Int. J. Geogr. Inf. Syst. Appl. Remote Sens.* **2013**, *4*, 12–25.
34. Chang, C.-Y.; Srinivasan, K.; Chen, M.-C.; Chen, S.-J. SVM-Enabled Intelligent Genetic Algorithmic Model for Realizing Efficient Universal Feature Selection in Breast Cyst Image Acquired via Ultrasound Sensing Systems. *Sensors* **2020**, *20*, 432. [CrossRef]
35. Baghaie, A.; Yu, Z. An optimization method for slice interpolation of medical images. *arXiv* **2014**, arXiv:1402.0936.
36. Chenevert, T.L.; Malyarenko, D.I.; Newitt, D.; Li, X.; Jayatilake, M.; Tudorica, A.; Fedorov, A.; Kikinis, R.; Liu, T.T.; Muzi, M.; et al. Errors in quantitative image analysis due to platform-dependent image scaling. *Trans. Oncol.* **2014**, *7*, 65–71. [CrossRef]
37. Rousseeuw, P.J.J. A graphical aid to the interpretation and validation of cluster analysis. *J. Comput. Appl. Math.* **1987**, *20*, 53–65. [CrossRef]
38. Del Re, E.C.; Gao, Y.; Eckbo, R.; Petryshen, T.L.; Blokland, G.A.M.; Seidman, L.J.; Konishi, J.; Goldstein, J.M.; McCarley, R.W.; Shenton, M.E.; et al. A New MRI Masking Technique Based on Multi-Atlas Brain Segmentation in Controls and Schizophrenia: A Rapid and Viable Alternative to Manual Masking. *J. Neuroimaging* **2016**, *26*, 28–36. [CrossRef]
39. Russakoff, D.B.; Tomasi, C.; Rohlfing, T.; Maurer, C.R. Image similarity using mutual information of regions. In Proceedings of the European Conference on Computer Vision, Prague, Czech Republic, 11 May 2004; pp. 596–607. [CrossRef]
40. Chaddad, A.; Tanougast, C. Quantitative evaluation of robust skull stripping and tumor detection applied to axial MR images. *Brain Inf.* **2016**, *3*, 53–61. [CrossRef] [PubMed]
41. Rajinikanth, V.; Satapathy, S.C.; Fernandes, S.L.; Nachiappan, S. Entropy-based segmentation of tumor from brain MR images–A study with teaching learning-based optimization. *Pattern Recognit. Lett.* **2016**, *94*, 87–94. [CrossRef]
42. Thanaraj, P.; Parvathavarthini, B. Multichannel interictal spike activity detection using time–frequency entropy measure. *Australas. Phys. Eng. Sci. Med.* **2017**, *40*, 413–425. [CrossRef] [PubMed]

43. Rajinikanth, V.; Dey, N.; Satapathy, S.C.; Ashour, A.S. An approach to examine magnetic resonance angiography based on Tsallis entropy and deformable snake model. *Future Gener. Future Comput. Syst.* **2018**, *85*, 160–172. [CrossRef]
44. Roopini, I.T.; Vasanthi, M.; Rajinikanth, V.; Rekha, M.; Sangeetha, M. Segmentation of tumour from brain MRI using fuzzy entropy and distance regularised level set. In Proceedings of the Computational Signal Processing and Analysis, Singapore, 3 April 2018; pp. 297–304.
45. Rajinikanth, V.; Satapathy, S.C.; Dey, N.; Vijayarajan, R. DWT-PCA Image fusion technique to improve segmentation accuracy in brain tumour analysis. In *Microelectronics, Electromagnetics and Telecommunications*; Springer: Singapore, 2018; pp. 453–462. Available online: https://link.springer.com/chapter/10.1007/978-981-10-7329-8_46 (accessed on 5 February 2020). [CrossRef]
46. Krishnan, P.T.; Balasubramanian, P.; Krishnan, C. Segmentation of brain regions by integrating meta heuristic multilevel threshold with markov random field. *Curr. Med. Imaging* **2016**, *12*, 4–12. [CrossRef]
47. Chen, Y.H.; Chang, C.C.; Lin, C.C.; Hsu, C.Y. Content-based color image retrieval using block truncation coding based on binary ant colony optimization. *Symmetry* **2019**, *11*, 21. [CrossRef]
48. Kalaiselvi, T.; Selvi, S.K. Investigation of Image Processing Techniques in MRI Based Medical Image Analysis Methods and Validation Metrics for Brain Tumor. *Curr. Med. Image Rev.* **2018**, *14*, 489–505. [CrossRef]
49. Rajinikanth, V.; Thanaraj, K.P.; Satapathy, S.C.; Fernandes, S.L.; Dey, N. Shannon's entropy and watershed algorithm based technique to inspect ischemic stroke wound. In *Smart Intelligent Computing and Applications*; Springer: Singapore, 2019; pp. 23–31.
50. Suresh, K.; Sakthi, U. Object Tracking based 3d Modelling and Quantification of Abnormal Contours in Brain MRI DICOM Study. *J. Eng. Sci. Technol.* **2019**, *14*, 2098–2115.
51. Shanthakumar, P.; Ganesh Kumar, P. Computer aided brain tumor detection system using watershed segmentation techniques. *Int. J. Image Syst. Technol.* **2015**, *25*, 297–301. [CrossRef]
52. Chan, T.F.; Vese, L.A. Active contours without edges. *IEEE Trans. Image Proc.* **2001**, *10*, 266–277. [CrossRef]
53. Memon, F.; Unar, M.A.; Memon, S. Image quality assessment for performance evaluation of focus measure operators. *arXiv* **2016**, arXiv:1604.00546.
54. Moga, A.N.; Gabbouj, M. Parallel marker-based image segmentation with watershed transformation. *J. Parallel Distrib. Comput.* **1998**, *51*, 27–45. [CrossRef]
55. Suresh, K.; Sakthi, U. A soft-computing based hybrid tool to extract the tumour section from brain MRI. *Multimed. Tools Appl.* **2019**, 1–5. [CrossRef]
56. Raja, N.S.; Fernandes, S.L.; Dey, N.; Satapathy, S.C.; Rajinikanth, V. Contrast enhanced medical MRI evaluation using Tsallis entropy and region growing segmentation. *J. Ambient Intel. Hum. Comput.* **2018**, 1–12. [CrossRef]
57. Chang, C.Y.; Srinivasan, K.; Hu, H.Y.; Tsai, Y.S.; Sharma, V.; Agarwal, P. SFFS–SVM based prostate carcinoma diagnosis in DCE-MRI via ACM segmentation. *Multidim. Syst. Sign. Process.* **2019**, 1–22. [CrossRef]
58. Kathiravan, S.; Kanakaraj, J. A Review of Magnetic Resonance Imaging Techniques. *Smart Comput. Rev.* **2013**, *3*, 358–366. [CrossRef]
59. Hua, K.; Dai, B.; Srinivasan, K.; Hsu, Y.H.; Sharma, V. A hybrid NSCT domain image watermarking scheme. *J. Image Video Proc.* **2017**, *2017*, 10. [CrossRef]
60. Srinivasan, K.; Sharma, V.; Jayakody, D.N.K.; Vincent, D.R. D-ConvNet: Deep learning model for enhancement of brain MR images. In Proceedings of the Basic & Clinical Pharmacology & Toxicology, Hoboken, NJ, USA, 23–24 December 2018; Volume 124, pp. 3–4. [CrossRef]
61. Srinivasan, K.; Ankur, A.; Sharma, A. Super-resolution of Magnetic Resonance Images using deep Convolutional Neural Networks. In Proceedings of the 2017 IEEE International Conference on Consumer Electronics—Taiwan (ICCE-TW), Taipei, China, 12–14 June 2017; pp. 41–42. [CrossRef]
62. Kathiravan, S.; Kanakaraj, J. A Review on Potential Issues and Challenges in MR Imaging. *Sci. World J.* **2013**, *2013*, 10. [CrossRef]
63. Srinivasan, K.; Kanakaraj, J. A Study on Super-Resolution Image Reconstruction Techniques. *Comput. Eng. Intell. Syst.* **2011**, *2*, 222–227.

64. Srinivasan, K.; Kanakaraj, J. An Overview of SR Techniques Applied to Images, Videos and Magnetic Resonance Images. *Smart Comput. Rev.* **2014**, *4*, 181–201. [CrossRef]
65. Hua, K.-L.; Trang, H.T.; Srinivasan, K.; Chen, Y.-Y.; Chen, C.-H.; Sharma, V.; Zomaya, A.Y. Reduction of Artefacts in JPEG-XR Compressed Images. *Sensors* **2019**, *19*, 1214. [CrossRef] [PubMed]

Article

Automated Volume Status Assessment Using Inferior Vena Cava Pulsatility

Luca Mesin [1,*], Silvestro Roatta [2], Paolo Pasquero [3] and Massimo Porta [3]

[1] Mathematical Biology and Physiology, Department of Electronics and Telecommunications, Politecnico di Torino, 10129 Turin, Italy

[2] Integrative Physiology Lab, Department of Neuroscience, Universitá di Torino, 10125 Turin, Italy; silvestro.roatta@unito.it

[3] Department of Medical Sciences, Universitá di Torino, 10126 Turin, Italy; pasquerop@gmail.com (P.P.); massimo.porta@unito.it (M.P.)

* Correspondence: luca.mesin@polito.it; Tel.: +39-011-090-4085

Received: 14 September 2020; Accepted: 7 October 2020; Published: 13 October 2020

Abstract: Assessment of volume status is important to correctly plan the treatment of patients admitted and managed by cardiology, emergency and internal medicine departments. Non-invasive assessment of volume status by echography of the inferior vena cava (IVC) is a promising possibility, but its clinical use is limited by poor reproducibility of current standard procedures. We have developed new algorithms to extract reliable information from non-invasive IVC monitoring by ultrasound (US) imaging. Both long and short axis US B-mode video-clips were taken from 50 patients, in either hypo-, eu-, or hyper-volemic conditions. The video-clips were processed to extract static and dynamic indexes characterizing the IVC behaviour. Different binary tree models (BTM) were developed to identify patient conditions on the basis of those indexes. The best classifier was a BTM using IVC pulsatility indexes as input features. Its accuracy (78.0% when tested with a leave-one-out approach) is superior to that achieved using indexes measured by the standard clinical method from M-mode US recordings. These results were obtained with patients in conditions of normal respiratory function and cardiac rhythm. Further studies are necessary to extend this approach to patients with more complex cardio-respiratory conditions.

Keywords: inferior vena cava; ultrasound imaging; binary tree model; pulsatility; fluid volume assessment

---

1. Introduction

The intravascular volume status (i.e., the extent of vascular filling) is a relevant cardiovascular parameter related to the cardiac preload (i.e., the stretch of cardiac tissue in relaxed conditions), which in turn affects cardiac output and arterial blood pressure. Its assessment in critically ill patients is essential to establish and carefully balance the appropriate fluid therapy, whereby fluid supplementation may favor cardiac efficiency, but also increase the rate of complications and mortality [1–3]. Various pathological conditions are characterized by alteration of the volume status, e.g., heart failure causes overload while dehydration leads to volume depletion.

The non invasive evaluation of the volume status is very important, as it allows to inspect a patient in emergency conditions or during the follow-up. An approximate assessment of the volemic condition can be obtained from US imaging of the IVC [4,5]. In fact, the pulsatility of the IVC was found to correlate with the intravascular fluid volume [6]. Moreover, it can be useful for the non-invasive estimation of the central venous pressure [7–9].

However, this non-invasive method has shown limitations [10,11]. Important problems are due to the lack of standardization [12] and to the subjectivity of the measurement [13]. Indeed, both B-mode and M-mode US scans have been used, followed by a subjective identification of the IVC maximal and minimal diameters [6,14]. Specifically, IVC pulsatility can be expressed in terms of the caval index (CI), defined as the variation of the vessel diameter during a respiratory cycle relative to the maximum diameter: minimum and maximum diameters are measured by the operator from a B-mode video or an M-mode trace considering inspiration and expiration, respectively. This clinical approach is not standardized [12,15] (e.g., either long [16] or short axis [17] visualizations are used), and is operator-dependent [13] and prone to measurement errors, e.g., due to movements [18] and non-uniform pulsatility of IVC [15]. In particular, the M-mode registration allows to visualize a section of the vein over time at high frequency along a fixed direction in space [19]. As the IVC moves during respiration, the M-mode approach fails to constantly refer to the same section of the vein. On the other hand, measuring diameters from a B-mode video requires that the operator chooses the frames corresponding to the end of inspiration and expiration, in addition to selecting the sections along which to estimate the diameters (which could be different, if the operator does not compensate for IVC movement).

Moreover, IVC pulsatility may vary considerably in different portions of the vein [15,20], so that a single measurement taken by a manual approach provides limited information.

Some confounding factors have also been documented in case of specific pathologies, e.g., to the respiratory system [21,22] or the heart [23,24]. Indeed, breathing and heartbeats provide the main stimulations affecting IVC pulsatility. Separating the effects of the two contributions could possibly help to counteract these limitations [13,25].

A further problem consists in the shape of the IVC: in different sections along the longitudinal view, the IVC can exhibit very different diameters, as in the case of a saber profile [15,26], and different pulsatilities [20]; the IVC cross-section can be irregular, much different from a circle or an ellipsoid (especially in the case of hypo-volemia), with a large variation of pulsatility in different directions [27,28]. Thus, investigating IVC both in long and short axis views and averaging across different sections or directions could be important to better characterize the vessel and its respirophasic dynamics.

We have carried out a series of studies trying to overcome some of the limitations of the US assessment of IVC. Specifically, we have developed an automated method to track the movements and estimate the borders of the IVC from US video-clips [19,20,28,29]. Data derived from both longitudinal (long axis) and transverse (short axis) sections of the vessel can be processed. The algorithms extract information from either a whole tract of longitudinal section of the vein or a transverse section, respectively. This way, the overall pulsatility can be estimated. Moreover, the border of the vein is found for each frame of the video-clip, so that the vessel pulsations can be investigated over time, obtaining time series which can be processed to estimate further indexes characterizing pulsatility, e.g., induced by either respiration or cardiac stimulation only. Preliminary results indicate that IVC average size and global pulsatility and its respiratory and cardiac components estimated in long axis are strongly related to the right atrial pressure (in contrast with size and pulsatility estimated by standard clinical approaches [30]) and are useful for its non-invasive estimation [25,31].

However, beyond these promising correlations, there is still no stable criterion capable of recognizing pathological problems in the volume status. With this study, we face this aspect and try to propose a classification method based on an automated processing. Specifically, US B-mode video-clips of IVC from long and short axis have been acquired from patients with different volume status. They have been processed, tracking the movements and estimating the border of the vein. Then, indexes characterizing IVC size and pulsatility have been automatically extracted and used to build a classifier able to discriminate patients in either hypo-, eu-, or hyper-volemic conditions.

2. Methods

2.1. US Video-Clip Processing

The physical dimension of a pixel was determined for each video-clip as a preliminary step, by scanning automatically a graduated length scale present in the frames. Then, the user selected some parameters (e.g., concerning the portion of frame to be processed, points to be tracked, etc.) needed by two algorithms (implemented in MATLAB R2018a, The Mathworks) that processed the B-mode US video-clips of the IVC in either long or short axis (notice that, as a preliminary interaction with the user is needed to process the US videos, we refer to our processing algorithms as semi-automated). Figure 1 shows an example of IVC border automated delineation obtained by those algorithms, described below. Once obtained the IVC borders in either of the two views, the mean diameter and pulsatility indexes (defined below) were estimated.

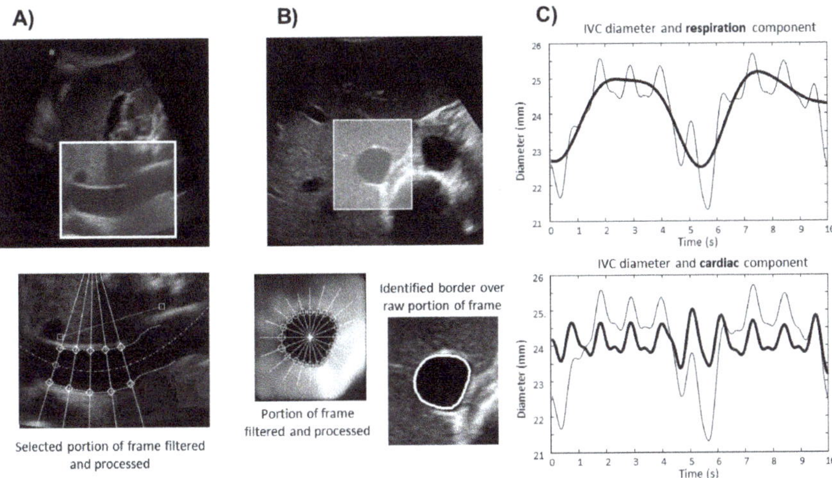

Figure 1. (**A**) Example of single frame of a video-clip of the IVC in long axis and result of processing (5 equidistant sections in direction orthogonal to the IVC midline are considered). (**B**) Example of frame in short axis and processing (median filter is applied on the bottom left figure, different rays are originated from the centre of the vein and their intersections with the border of the vessel are computed; estimated IVC border indicated in the bottom right image). (**C**) Representative example of IVC diameter over time (the average of the diameters of the case shown in A is considered, only for representation purposes): from the time series, respiratory and cardiac components (the latter added to the average diameter, for clearer representation) are extracted by specific filters.

2.1.1. Identification of IVC Borders in Long Axis

The algorithm proposed in [20] (and already applied in [13,25,31]) was used. In the first frame of the clip, the user located the vein and the region of interest, indicating two reference points to be tracked to compensate for IVC movements and deformations, the leftmost and rightmost lines to be considered and the location of the borders of the vein in the leftmost line. In optimal conditions, the available tract was between the confluence of the hepatic veins into the IVC and the caudate lobe of the liver. Each frame was first pre-processed with a 2D median filter (neighborhoods of 9 × 9 pixels). Then, the software uniformly distributed 21 lines in the region of interest and identified the borders of the vein along these lines (as a jump of the US intensity along them). For each subsequent frame, the location and direction of those lines were updated based on the estimated movements of the reference points from the previous one.

Once obtained the superior and inferior borders of the vein, the software computed the IVC midline and distributed uniformly 5 points along it. For our specific application, the extension of the midline was considered only from the 20% to the 80% of its length, thus excluding the edges. Sections orthogonal to the IVC midline passing from each of these 5 points were considered and the IVC pulsatility was estimated for each of them.

2.1.2. Identification of IVC Borders in Short Axis

The algorithm proposed in [28] (and already used in [29]) was employed. The user was asked to indicate the centre of the IVC and to draw a rectangle enclosing it in the first frame of the video-clip. Subsequent frames were cropped in a rectangular region with the same dimension, centered on the IVC estimated on the previous frame.

The image was converted in gray-scale, contrast enhanced using histogram equalization and processed with a 2D median filter (neighborhoods of 11×11 pixels). The outline of the vein was then estimated by the algorithm. Twenty rays were defined, originating from the centre of the considered rectangular portion of image and sampling uniformly the directions around it. For each ray, the intensity of the image along it was estimated by cubic interpolation. The border of the vein was identified as an abrupt increase of the intensity (from the lumen to the outside tissues).

Once the 20 border points were found, their coordinates were low pass filtered (Butterworth non-causal, zero-phase IIR filter of order 4 with cut-off at 0.3) to get a smooth boundary of the vein. Furthermore, the maximum variation of the length of a ray was imposed to be 5 pixels; the rays which overcame such a threshold were removed and substituted by a quadratic interpolation of the 4 closest neighboring border points.

2.1.3. IVC Indexes

The mean diameter was estimated averaging both across different sections (i.e., 5 sections in long axis and 10 diameters corresponding to the 20 rays in short axis) and time (i.e., considering the frames of the video-clips).

Pulsatility was measured in terms of the CI

$$\mathrm{CI} = \frac{\max_t (D(t)) - \min_t (D(t))}{\max_t (D(t))} \tag{1}$$

where D indicates the dimension over the time variable t of IVC, expressed either as diameter or equivalent diameter (proportional to the square root of the area [28]), in the long and short axis, respectively, and max/min indicate local extrema. Local maxima and minima were computed for each respiratory cycle. A CI accounting for the overall pulsatility was obtained by averaging the estimations across different respiratory cycles and different sections (the latter, only in the case of the long axis approach).

Additional indexes were also estimated by decomposing the time series reflecting IVC pulsations into low and high frequency components (below 0.4 Hz and above 0.8 Hz, respectively), assumed to reflect the stimulations induced by either respiration or heartbeats, respectively (both filters were 4th order Butterworth, used twice, once with time reversed, to remove phase distortion and delay). From these filtered time series, applying again the definition of CI (1) on local maxima and minima, the respiratory caval index (RCI) and the cardiac caval index (CCI) were obtained. Stable estimations of both indexes were computed by averaging across either respiratory cycles or heartbeats (and on the 5 sections, in the case of the long axis).

An example of time series extracted from a video-clip is given in Figure 1C.

IVC was also investigated by standard manual measurements, in both long and short axis, in M-mode. Stable estimations of the minimum and maximum IVC diameter were obtained by averaging across more measurements (up to 3). Then, the maximum and minimum diameters were used to compute the CI and the average IVC diameter (defined as the mean of the two diameters).

2.2. Experimental Data

Inclusion criteria were the presence of pathological conditions in the Emergency Department and in the Department of Medicine resulting in overload (heart failure) or volume depletion (dehydration or moderate bleeding). As a control group, patients without the previous conditions were selected. Exclusion criteria were chronic obstructive pulmonary disease, pulmonary hypertension, interstitial disease or thromboembolism, tension pneumothorax, cirrhosis and/or ascitic effusion, serum creatinine >3 mg/dl, constrictive pericarditis and cardiac tamponade. Fifty patients were included in the study. They were selected from a database of 69 patients (Table 1). On the basis of clinical considerations (based upon physical examination, laboratory data and imaging), each patient was associated to one of the following classes:

1. hypo-volemic condition (20 subjects);
2. eu-volemic condition (24 subjects);
3. hyper-volemic condition (25 subjects).

US B-mode video-clips of about 15 s were recorded bedside in spontaneous breathing, with subxifoideal approach, using a MyLab Seven system (Esaote, Genova, Italy; frame rate 30 Hz, 256 gray levels) equipped with a convex 2–5 MHz probe. M-mode scans were also recorded to allow for standard manual measurements.

According to the Declaration of Helsinki, subjects provided written informed consent for the collection of data and subsequent analysis. The study was approved by the local Ethics Committee.

The data included here were only those for which both video-clips, recorded along either long or short axis, could be reliably processed. They were 20 from patients in overload, 19 controls and 11 patients with volume depletion (notice that video-clips of patients with volume depletion are more difficult to be processed, due to the small dimension of the IVC, which could even collapse in some frames, hindering proper processing).

Figure 2 shows examples of patients in the 3 classes.

Table 1. Number of patients included in different groups (with indication of the entire database and of the patients for which successful processing of both long and short axis ultrasound videos was achieved).

	Hypo-Volemic	Eu-Volemic	Hyper-Volemic
Database	20	24	25
Successful processing	11	19	20
Rate of successful processing	55.0%	79.2%	80.0%

2.3. Automated Identification of the Volemic Status

Three different classification approaches were fit to our dataset, as a preliminary step: the error-correcting output codes (ECOC) model, using support vector machines (SVM [32]) for binary one-to-one classifications [33,34]; the Naive Bayes classifier (estimating data distributions using smoothed densities with normal kernel) [35]; the BTM [35]. For each approach, different models were fit to our data, considering all possible combinations of input features (detailed below). The performances of different classifiers were compared in terms of a 10-fold cross-validation test, which allowed to select the best input features and classification approach. Then, the selected classifier was tested by a leave-one-out approach and, finally, trained on the entire dataset, to provide an ultimate prediction model. In the following, we will focus only on the BTM, as best results were obtained using this approach.

Different BTMs were fit to our multi-class classification problem (including 3 classes), selecting the simplest one (i.e., with minimum dimension) with best performances. A BTM iteratively splits the dataset in two groups, after comparing an index with a threshold (Gini's diversity index was used as splitting criterion). Thus, it is built by choosing the optimal number of splittings, the specific index to be considered for each binary separation and selecting the threshold value for each splitting. Different BTMs were developed considering all possible combinations of input indexes (exhaustive search): all possible choices of a single index, all pairs, triplets, ... until using all indexes.

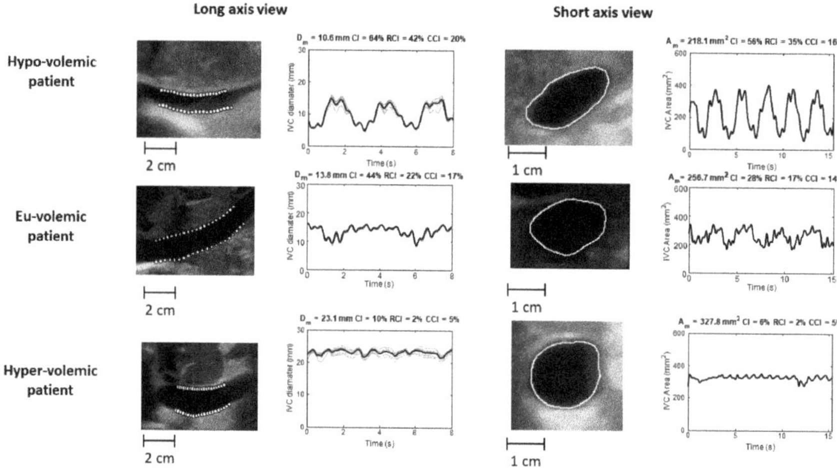

Figure 2. Examples of data from patients in either hypo-, eu- or hyper-volemic conditions. The first frames of the long and short axis scans are shown (left and right, respectively), together with the IVC boundaries identified by the algorithm. Time series are also shown for the diameters in 5 sections of the IVC (in gray, with superimposed the mean diameter in black) and for the IVC area estimated from the long and short axis scans, respectively. In the case of long axis scans, pulsatility indexes were computed as averages of estimations from each of the 5 sections; in the case of short axis scans, they were computed from the equivalent diameter, proportional to the square root of the IVC cross-section area. D_m: mean diameter; A_m: mean area; CI: caval index; RCI: respiratory caval index; CCI: cardiac caval index.

Different sets of indexes were used, considering the possibility of either employing the semi-automated processing or not (so that in the latter case only manual measurements were considered).

The set of indexes obtained by semi-automated video processing was the following:

1. mean diameter of IVC in long axis;
2. CI in long axis;
3. RCI in long axis;
4. CCI in long axis;

5. equivalent diameter of IVC in short axis;
6. CI in short axis;
7. RCI in short axis;
8. CCI in short axis.

The set of indexes obtained by standard manual measurements was the following:

1. diameter of IVC in long axis;
2. CI in long axis;
3. diameter of IVC in short axis;
4. CI in short axis.

For each set of indexes, different BTMs were developed using all possible combinations of features taken from it and the one with highest performance was selected (thus, they were 255 and 15, for the first and second features set, respectively). Specifically, the best categorical predictor split was chosen from all possible combinations of choices. As mentioned above, the models were cross-validated considering 10 folds. The order of the data was random, so that the three categories of patients had a similar representation in each fold (however, they could not be equally represented; this problem is emphasized by the small size of our dataset). The one providing minimum average root mean squared regression error (or loss) on the validation sets was then selected. This specific model was then tested by a leave-one-out approach, to reduce the bias in error estimation (considering our small dataset) [36].

3. Results

Indexes characterizing the IVC were extracted from long and short axis views by either semi-automated processing or manual estimation (performed in M-mode). Then, they were used to classify patients. As using indexes extracted with the automated processing resulted in better performances, figures and tables shown below refer to those data, indicating in the text some performance indexes of the best BTM developed using the set of indexes obtained by standard manual measurements.

Figure 3 shows the BTM selected as the classifier with best performances on our dataset. The shown BTM was trained on the entire dataset, including the best input features selected by the cross-validation test (described in Section 2.3), where minimum loss was obtained (equal to 0.26; the loss of the best classifiers using either ECOC or Naive Bayes models was 0.28).

Two pulsatility indexes are included: CCI in long axis and CI in short axis. The same loss was obtained by other 4 BTMs: the one with minimum number of input features was selected. The CCI in long axis was included in 4 of these BTMs with minimum loss; the CI in short axis was included in 2 of them. Another feature which was often included was the RCI in short axis, which was used in 3 among the 5 BTMs with minimum loss. In the case in which standard manual measurements were employed, the best BTM was unique, it had a loss of 0.28 and included two indexes: IVC diameter estimated in long axis and CI in short axis.

Distributions of the indexes are shown in Figure 4. The mean Fisher ratios (FR, considering all 3 binary comparisons) of the indexes selected by the best BTM between those estimated by semi-automated processing are among the highest. However, they have not the highest FRs: indeed, the best discrimination in terms of average FR is provided by the mean diameter estimated from the long axis view. This indicates that the selected indexes are those that are both informative and not much redundant, allowing a peak in performance of the classifier using them as inputs. Notice also that the FR is an index of linear discrimination, whereas the adopted classifier allows for nonlinear separation.

It is interesting to see that the indexes estimated manually have even higher FRs, indicating a better linear discrimination of the patients. The two indexes with highest FRs are those selected by the best

BTM using only indexes measured manually. However, the semi-automated processing allows to extract additional information: specifically, the two pulsatility indexes RCI and CCI reflect the effect of different stimulations (respiration and heartbeat, respectively). This further information (and specifically that coming from the CCI) allows the BTM from automated processing to get better performances than the one developed on the basis of the set of manually estimated indexes.

The confusion matrix of the best BTM shown in Figure 3 is given in Table 2. Notice that all hypo-volemic patients were correctly identified. A few eu-volemic and hyper-volemic subjects were misclassified. No hyper-volemic patient was confused as hypo-volemic or vice-versa. Common performance indexes are the followings: mean sensitivity 90.0% (86.0% for the BTM built using the manually estimated indexes); mean specificity 95.0% (91.9% with manual indexes); positive predictive value 90.0% (86.2% with manual indexes); negative predictive value 94.2% (91.8% with manual indexes); mean accuracy 92.9% (89.8% with manual indexes).

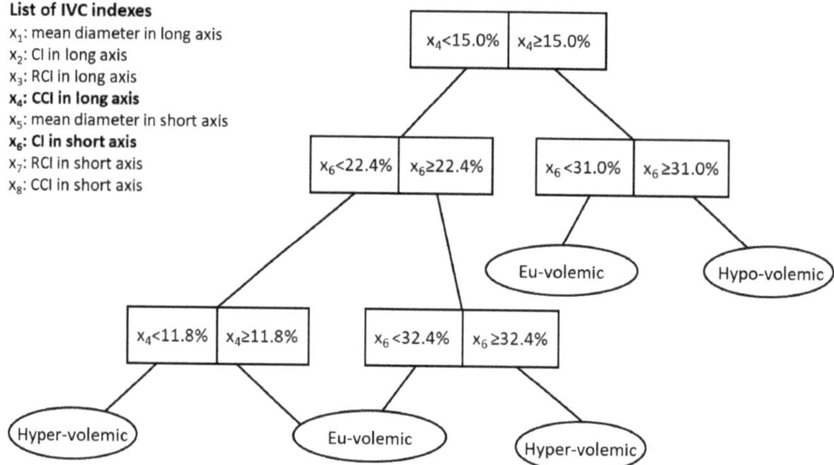

Figure 3. BTM with best performances in fitting our data. The list of tested indexes (all estimated by automated processing) is also provided, with indication (in bold) of those selected by the BTM.

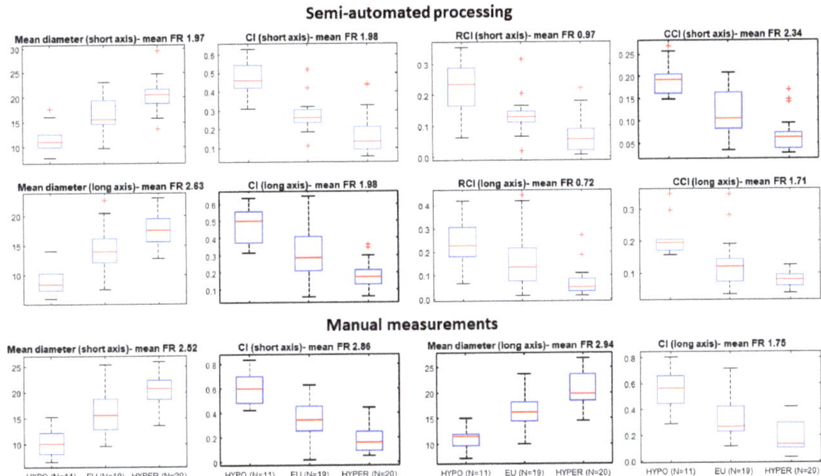

Figure 4. Distribution of the considered IVC indexes from patients with different volemic conditions. The FR (ratio between squared difference of means and sum of variances, computed for all 3 binary comparisons and averaged) is indicated, as an index of linear discrimination. The indexes selected by the best BTMs (those using either semi-automated or manually estimation approach) are emphasized.

Notice that these performances were obtained using the entire dataset to train our model. As some misclassifications were obtained, we deduce that some information is still missing and/or the features extracted by our processing contain some residual noise. To get a more faithful indication of performances, a leave-one-out test was performed (i.e., the best features selected before were kept, but each sample was excluded in turn from the training set and used for testing). The confusion matrix in Table 3 was obtained. Some degradation of the performance can be observed, especially in the discrimination of the control and hyper-volemic groups. The following performance indexes were achieved: mean sensitivity 70.0% (66.0% for the BTM built using the manually estimated indexes and tested by a leave-one-out approach); mean specificity 83.2% (80.4% with manual indexes); positive predictive value 70.0% (65.1% with manual indexes); negative predictive value 82.1% (80.5% with manual indexes); mean accuracy 78.0% (75.3% with manual indexes).

Table 2. Confusion matrix of the best binary tree model classifying the volemic status, shown in Figure 3 (for comparison, the best error-correcting output codes and Naive Bayes classifiers trained on the entire dataset show a predictive value of 78% and 86%, respectively).

Predicted	Target Score			Predictive
Class	1: Hypo	2: Eu	3: Hyper	Value
1	11 (22.0%)	2 (4.0%)	0	84.6%
2	0	15 (30.0%)	1 (2.0%)	93.8%
3	0	2 (4.0%)	19 (38.0%)	90.5%
True rate	100%	78.9%	95.0%	**90.0%**

Table 3. Confusion matrix obtained by testing the best binary tree model with a leave-one-out approach.

Predicted	Target Score			Predictive
Class	1: Hypo	2: Eu	3: Hyper	Value
1	10 (20.0%)	3 (6.0%)	0	76.9%
2	1 (2.0%)	12 (24.0%)	7 (14.0%)	60.0%
3	0	4 (8.0%)	13 (26.0%)	76.5%
True rate	90.9%	63.2%	65.0%	**70.0%**

4. Discussion

The accurate assessment of the volume status is of relevance for a high percentage of patients who either access the emergency room or enter the medical wards. The development of new standardized clinical procedures and the support of automatic algorithms can help to correctly plan the treatment and monitor the follow-up.

We have developed two algorithms that help standardizing the assessment of IVC pulsatility. They process B-mode video-clips, allowing to compensate for either longitudinal or transverse respirophasic movements and to delineate the vessel edges in an entire region (either a tract of longitudinal section or a cross-section). As detailed in the Methods section, the IVC indexes extracted are not related to a single diameter along the IVC, but reflect the average size and pulsatility, calculated over the whole portion of considered IVC length (in long axis) or over the entire IVC cross-section (in short axis). Thus, the considered IVC indexes reflect the overall behaviour of the investigated regions (both static and dynamic behaviour, reflected by the size and pulsatility of the IVC, respectively). We have already documented for the long axis scans that this approach is more reliable and repeatable than standard clinical assessment [13,19,25]. Moreover, we have shown that the IVC in a short axis view can pulsate differently along different directions [28], so that an average indication of cross-sectional pulsatility is preferable and less subjective than referring to an arbitrarily chosen diameter.

Here, we have built BTMs using those indexes estimated by our algorithms to assess automatically the volemic conditions of patients. The indexes used by the best BTM reflect IVC pulsatility. Referring to Figure 3, the joint integration of information from CCI from long axis US scans and CI in short axis allows to identify the different conditions, with an accuracy of 78% in a leave-one-out test (larger than what could be achieved with the best classifier using only manual indexes). Notice that CCI is an index that was introduced recently [13,25,37,38] and whose estimation is expected to be stable, as the heartbeats are much less variable than respiratory cycles, mainly affecting the measurements of CI and RCI. Other 4 BTMs achieved the same loss in cross-validation as the best one (which was chosen because it had the smallest dimension). This can be interpreted as a consequence of the redundancy included in the pulsatility indexes, whereby one index may be obtained from a combination of the others. This result may also descend from the small sample size, which does not allow to appreciate fine differences in performance among models with high classification rates. Hence, this should be considered as a pilot study. Augmenting the numerosity of the sample would be important to get a more stable estimation of the classification model.

We have compared the classification performances of the above mentioned fully automated method (based on indexes extracted by processing US B-mode video-clips), with a BTM using indexes measured manually, with M-mode scans along an US ray selected either from a longitudinal or a transverse view of the IVC. It is interesting to notice that the indexes measured manually allowed in general to get a better linear discrimination of the volemic conditions (measured in terms of the average Fisher ratios comparing all pairs of groups). However, the automated processing allowed to extract more indexes describing IVC and the final best classifier showed better performances than that obtained using manual measurements. In particular, additional information on IVC pulsatility induced by either respiratory cycles or heartbeats was available and CCI (from long axis view) was selected by the best BTM. We deduce that this index includes additional/not redundant information that, together with other characterizations of IVC pulsatility (provided by the CI in short axis, in the best BTM), can be useful to disentangle the complex/nonlinear relation between IVC dynamics and volume status of the patient.

It must be underlined that the present results were obtained from a selected group of patients in which pathologies specifically affecting the respiratory system were excluded. Moreover, we expect that the selection of CCI as optimal feature depends also on the regular hearth rhythm shown by the patients included in our dataset; in the case of arrhythmia (typically due to atrial fibrillation, not shown by our data sample), a reliable estimation of this important parameter would be hindered. Thus, the application of our classification approach to different patient populations could result in different selections of parameters and thresholds. Nevertheless, as an effort to overcome the subjectivity of the measurement, our approach (but, probably, not the classification model) remains valid and worth to be investigated and extended to other patients groups.

Another limitation of the method is the need to rely on good quality imaging. Indeed, only 87% and 77% of long and short axis video-clips were properly processed, respectively, so that only 72% of our patients could be included in this study (as the processing of both recordings was required). Improvements could be obtained by adopting higher level US machines or by more effective image processing. We are currently trying to optimize our algorithms in order to process US recordings in real time, providing a feedback to the operator. We expect that this could help in getting successful processing in more US video-clips. Indeed, our present offline approach requires that the operator acquires data blindly, i.e., without knowing if the recorded video-clip will be adequate for processing. Instead, a real time software could guide the acquisition and indicate to the operator if there are problems in processing the data, in which case the operator could work at improving the quality of the imaging. This is exactly what happens in manual measurements: the operator may try different approaches and strategies to improve image quality until he is satisfied with the result and IVC measurement is made possible.

In summary, we have shown the joint application of long and short axis US views of the IVC, to assess the volume status of patients. The US videos have been automatically processed by multi-section and multi-directional algorithms, which track IVC movements and compute its size and pulsatility either over a longitudinal portion of the vessel or a cross-section, respectively. The IVC pulsations have been also split into two contributions, reflecting either the respiratory cycles or the heartbeats. The algorithms have been widely tested on healthy subjects in laboratory conditions in the past [13,19,20,28,29] and in a single clinical study, aimed at estimating right atrial pressure based on the analysis of IVC pulsatility [25,31]. Here, the different indexes were jointly applied in a clinical setting and used to solve the multiclass problem of discriminating patients with different volume status, showing better performance than when using manually measured indexes. Pulsatility indexes estimated from both long and short axis have been included in the best classification model, which supports the concept that they convey complementary information. Even considering the preliminary nature of these results (given the small sample size), the approach appears to be very promising. Extending the dataset and improving the processing algorithms (e.g., allowing real time interaction with the operator) may prospectively lead to obtain efficient systems for diagnostic support and follow-up.

5. Conclusions

The identification of the volume condition is important for the clinical management of many patients in the emergency room or in wards of general medicine and cardiology. We propose an automated approach for the classification of the volemic status, based on the processing of B-mode US video-clips of the IVC and on the extraction of pulsatility features. The presented results suggest that this approach may be useful to get more reliable clinical indication from the US monitoring of IVC. Investigation over a larger dataset will however be necessary to test the actual effectiveness of the proposed method. Moreover, our results hold true in conditions of normal respiratory function and cardiac rhythm. It is reasonable that our classifier will not apply to patients with more complex cardio-respiratory conditions; however, the same approach could be applied to develop models fitting their conditions.

6. Patents

An instrument implementing the algorithms for IVC delineation used in this paper was patented by Politecnico di Torino and Universitá di Torino (WO 2018/134726).

Author Contributions: Conceptualization, L.M., S.R., P.P. and M.P.; methodology, L.M.; software, L.M.; validation, L.M.; data preparation, P.P. and M.P.; investigation, L.M. and S.R.; writing–Original draft preparation, L.M.; writing–Review and editing, S.R., M.P., P.P.; visualization, L.M.; supervision, M.P. All authors have read and agreed to the published version of the manuscript.

Funding: This research was carried out as part of the project "Method and apparatus to characterise non-invasively images containing venous blood vessels", funded through the PoC Instrument initiative, implemented by LINKS, with the support of LIFFT, with funds from Campagnia di San Paolo.

Conflicts of Interest: The authors declare no conflict of interest.

Abbreviations

The following abbreviations are used in this manuscript:

BTM	Binary Tree Model
CI	Caval Index
CCI	Cardiac Caval Index
ECOC	Error-Correcting Output Codes
FR	Fisher ratio
IVC	Inferior Vena Cava
RCI	Respiratory Caval Index
SVM	Support Vector Machine
US	Ultrasound

References

1. Alsous, F.; Khamiees, M.; DeGirolamo, A.; Amoateng-Adjepong, Y.; Manthous, C. Negative fluid balance predicts survival in patients with septic shock: A retrospective pilot study. *Chest* **2000**, *117*, 1749–1754. [CrossRef] [PubMed]
2. Boyd, J.; Forbes, J.; Nakada, T.; Walley, K.; Russell, J. Fluid resuscitation in septic shock: A positive fluid balance and elevated central venous pressure are associated with increased mortality. *Crit. Care Med.* **2011**, *39*, 259–265. [CrossRef]
3. Murphy, C.; Schramm, G.; Doherty, J.; Reichley, R.; Gajic, O.; Afessa, B.; Micek, S.; Kollef, M. The importance of fluid management in acute lung injury secondary to septic shock. *Chest* **2009**, *136*, 102–109. [CrossRef]
4. Dipti, A.; Soucy, Z.; Surana, A.; Chandra, S. Role of inferior vena cava diameter in assessment of volume status: A meta-analysis. *Am. J. Emerg. Med.* **2012**, *30*, 1414–1419. [CrossRef] [PubMed]
5. Kelly, N.; Esteve, R.; Papadimos, T.J.; Sharpe, R.P.; Keeney, S.A.; DeQuevedo, R.; Portner, M.; Bahner, D.P.; Stawicki, S.P. Clinician–performed ultrasound in hemodynamic and cardiac assessment: A synopsis of current indications and limitations. *Eur. J. Trauma Emerg. Surg.* **2015**, *41*, 469–480. [CrossRef] [PubMed]

6. Yamanoolu, N.G.C.; Yamanoolu, A.; Parlak, Y.; Pynar, P.; Tosun, A.; Erkuran, B.; Aydynok, G.; Torlak, F. The role of inferior vena cava diameter in volume status monitoring; the best sonographic measurement method? *Am. J. Emerg. Med.* **2015**, *33*, 433–438.
7. Ciozda, W.; Kedan, I.; Kehl, D.W.; Zimmer, R.; Khandwalla, R.; Kimchi, A. The efficacy of sonographic measurement of inferior vena cava diameter as an estimate of central venous pressure. *Cardiovasc. Ultrasound* **2016**, *14*, 33. [CrossRef] [PubMed]
8. Nagdev, A.; Merchant, R.; Tirado-Gonzalez, A.; Sisson, C.; Murphy, M. Emergency department bedside ultrasonographic measurement of the caval index for noninvasive determination of low central venous pressure. *Ann. Emerg. Med.* **2010**, *55*, 290–295. [CrossRef]
9. Stawicki, S.P.; Braslow, B.M.; Panebianco, N.L.; Kirkpatrick, J.N.; Gracias, V.H.; Hayden, G.E.; Dean, A.J. Intensivist use of hand-carried ultrasonography to measure IVC collapsibility in estimating intravascular volume status: Correlations with CVP. *J. Am. Coll. Surg.* **2009**, *209*, 55–61. [CrossRef]
10. Gui, J.; Yang, Z.; Ou, B.; Xu, A.; Yang, F.; Chen, Q.; Jiang, L.; Tang, W. Is the Collapsibility Index of the Inferior Vena Cava an Accurate Predictor for the Early Detection of Intravascular Volume Change? *Shock* **2018**, *49*, 29–32. [CrossRef]
11. Long, E.; Oakley, E.; Duke, T.; Babl, F.E. Paediatric Research in Emergency Departments International Collaborative (PREDICT). Does Respiratory Variation in Inferior Vena Cava Diameter Predict Fluid Responsiveness: A Systematic Review and Meta-Analysis. *Shock* **2017**, *47*, 550–559. [CrossRef] [PubMed]
12. Zhang, Z.; Xu, X.; Ye, S.; Xu, L. Ultrasonographic measurement of the respiratory variation in the inferior vena cava diameter is predictive of fluid responsiveness in critically ill patients: Systematic review and meta-analysis. *Ultrasound Med. Biol.* **2014**, *40*, 845–853. [CrossRef] [PubMed]
13. Mesin, L.; Giovinazzo, T.; D'Alessandro, S.; Roatta, S.; Raviolo, A.; Chiacchiarini, F.; Porta, M.; Pasquero, P. Improved repeatability of the estimation of pulsatility of inferior vena cava. *Ultrasound Med. Biol.* **2019**, *45*, 2830–2843. [CrossRef] [PubMed]
14. Finnerty, N.M.; Panchal, A.R.; Boulger, C.; Vira, A.; Bischof, J.J.; Amick, C.; Way, D.P.; Bahner, D.P. Inferior Vena Cava Measurement with Ultrasound: What Is the Best View and Best Mode? *West. J. Emerg. Med.* **2017**, *18*, 496–501. [CrossRef] [PubMed]
15. Wallace, D.J.; Allison, M.; Stone, M.B. Inferior vena cava percentage collapse during respiration is affected by the sampling location: An ultrasound study in healthy volunteers. *Acad. Emerg. Med.* **2010**, *17*, 96–99. [CrossRef] [PubMed]
16. Barbier, C.; Loubieres, Y.; Schmit, C.; Hayon, J.; Ricome, J.L.; Jardin, F.; Vieillard-Baron, A. Respiratory changes in inferior vena cava diameter are helpful in predicting fluid responsiveness in ventilated septic patients. *Intensive Care Med.* **2004**, *30*, 1740–1746. [CrossRef]
17. Blehar, D.J.; Dickman, E.; Gaspari, R. Identification of congestive heart failure via respiratory variation of inferior vena cava diameter. *Am. J. Emerg. Med.* **2009**, *27*, 71–75. [CrossRef]
18. Blehar, D.; Resop, D.; Chin, B.; Dayno, M.; Gaspari, R. Inferior vena cava displacement during respirophasic ultrasound imaging. *Crit. Ultrasound J.* **2012**, *4*, 1–5. [CrossRef]
19. Mesin, L.; Pasquero, P.; Albani, S.; Porta, M.; Roatta, S. Semi-automated tracking and continuous monitoring of inferior vena cava diameter in simulated and experimental ultrasound imaging. *Ultrasound Med. Biol.* **2015**, *41*, 845–857. [CrossRef]
20. Mesin, L.; Pasquero, P.; Roatta, S. Tracking and Monitoring Pulsatility of a Portion of Inferior Vena Cava from Ultrasound Imaging in Long Axis. *Ultrasound Med. Biol.* **2019**, *45*, 1338–1343. [CrossRef]
21. Bouzat, P.; Walther, G.; Rupp, T.; Levy, P. Inferior Vena Cava Diameter May Be Misleading in Detecting Central Venous Pressure Elevation Induced by Acute Pulmonary Hypertension. *Am. J. Respir. Crit. Care Med.* **2014**, *190*, 233–235. [CrossRef] [PubMed]
22. Inocencio, M.; Childs, J.; Chilstrom, M.L.; Berona, K. Ultrasound Findings in Tension Pneumothorax: A Case Report. *J. Emerg. Med.* **2017**, *52*, e217–e220. [CrossRef] [PubMed]
23. Hjemdahl-Monsen, C.E.; Daniels, J.; Kaufman, D.; Stern, E.H.; Teichholdz, L.E.; Meltzer, R.S. Spontaneous contrast in the Inferior vena cava in a patient with constrictive Pericarditis. *J. Am. Coll. Cardiol.* **1984**, *4*, 165–167. [CrossRef]

24. Pellicori, P.; Carubelli, V.; Zhang, J.; Castiello, T.; Sherwi, N.; Clark, A.L.; Cleland, J.G.F. IVC diameter in patients with Chronic Heart Failure: Relationships and prognostic significance. *J. Am. Coll. Cardiol. Cardiovasc. Imaging* **2013**, *6*, 16–28. [CrossRef] [PubMed]
25. Mesin, L.; Albani, S.; Sinagra, G. Non-invasive Estimation of Right Atrial Pressure using the Pulsatility of Inferior Vena Cava. *Ultrasound Med. Biol.* **2019**, *45*, 1331–1337. [CrossRef]
26. Lichtenstein, D. *Inferior Vena Cava. General Ultrasound in the Critically Ill*; Springer: Berlin, Germany, 2005; Volume 23, p. 82.
27. Huguet, R.; Fard, D.; d'Humieres, T.; Brault-Meslin, O.; Faivre, L.; Nahory, L.; Dubois-Randé, J.; Ternacle, J.; Oliver, L.; Lim, P. Three-dimensional inferior vena cava for assessing central venous pressure in patients with cardiogenic shock. *J. Am. Soc. Echocardiogr.* **2018**, *31*, 1034–1043. [CrossRef]
28. Mesin, L.; Pasquero, P.; Roatta, S. Multi-directional assessment of Respiratory and Cardiac Pulsatility of the Inferior Vena Cava from Ultrasound Imaging in Short Axis. *Ultrasound Med. Biol.* **2020**, in press. [CrossRef]
29. Folino, A.; Benzo, M.; Pasquero, P.; Laguzzi, A.; Mesin, L.; Messere, A.; Porta, M.; Roatta, S. Vena Cava Responsiveness to Controlled Isovolumetric Respiratory Efforts. *J. Ultrasound Med.* **2017**, *36*, 10, 2113–2123. [CrossRef]
30. Magnino, C.; Omedé, P.; Avenatti, E.; Presutti, D.; Iannaccone, A.; Chiarlo, M.; Moretti, C.; Gaita, F.; Veglio, F.; Milan, A. RIGHT1 Investigators. Inaccuracy of right atrial pressure estimates through inferior vena cava indices. *Am. J. Cardiol.* **2017**, *120*, 1667–1673. [CrossRef]
31. Albani, S.; Pinamonti, B.; Giovinazzo, T.; de Scordilli, M.; Fabris, E.; Stolfo, D.; Perkan, A.; Gregorio, C.; Barbati, G.; Geri, P.; et al. Accuracy of right atrial pressure estimation using a multi-parameter approach derived from inferior vena cava semi-automated edge-tracking echocardiography: A pilot study in patients with cardiovascular disorders. *Int. J. Cardiovasc. Imaging* **2020**, *36*, 1213–1225. [CrossRef]
32. Bevilacqua, V.; Dimauro, G.; Marino, F.; Brunetti, A.; Cassano, F.; Maio, A.; Nasca, E.; Trotta, G.F.; Girardi, F.; Ostuni, A.; et al. A novel approach to evaluate blood parameters using computer vision techniques. In Proceedings of the IEEE International Symposium on Medical Measurements and Applications (MeMeA), Benevento, Italy, 15–18 May 2016; pp. 1–6.
33. Khurram, I.Q.; Lam, H.K.; Bo, X.; Gaoxiang, O.; Xunhe, Y. Classification of epilepsy using computational intelligence techniques. *CAAI Trans. Intell. Technol.* **2016**, *1*, 2, 137–149.
34. Furnkranz, J. Round Robin Classification. *J. Mach. Learn. Res.* **2002**, *2*, 721–747.
35. Trevor, H.; Tibshirani, R.; Friedman, J. *The Elements of Statistical Learning*; Springer Series in Statistics; Springer: New York, NY, USA, 2008.
36. Varma, S.; Simon, R. Bias in error estimation when using cross-validation for model selection. *BMC Bioinform.* **2006**, *7*, 91.
37. Nakamura, K.; Tomida, M.; Ando, T.; Sen, K.; Inokuchi, R.; Kobayashi, E.; Nakajima, S.; Sakuma, I.; Yahagi, N. Cardiac variation of inferior vena cava: New concept in the evaluation of intravascular blood volume. *J. Med. Ultrasound* **2013**, *40*, 205–209. [CrossRef] [PubMed]
38. Sonoo, T.; Nakamura, K.; Ando, T.; Sen, K.; Maeda, A.; Kobayashi, E.; Sakuma, I.; Doi, K.; Nakajima, S.; Yahagi, N. Prospective analysis of cardiac collapsibility of inferior vena cava using ultrasonography. *J. Crit. Care* **2015**, *30*, 945–948. [CrossRef] [PubMed]

MDPI
St. Alban-Anlage 66
4052 Basel
Switzerland
Tel. +41 61 683 77 34
Fax +41 61 302 89 18
www.mdpi.com

Electronics Editorial Office
E-mail: electronics@mdpi.com
www.mdpi.com/journal/electronics

www.ingramcontent.com/pod-product-compliance
Lightning Source LLC
LaVergne TN
LVHW070625170726
843515LV00004B/182

9783036503462